# COMPOSITE PANELS/PLATES

# Composite
## Panels/Plates

### ANALYSIS AND DESIGN

## Rafaat M. Hussein, Ph.D.
UNIVERSITY OF EVANSVILLE
EVANSVILLE, INDIANA

TECHNOMIC
PUBLISHING CO., INC

LANCASTER · BASEL

*Published in the Western Hemisphere by*
Technomic Publishing Company, Inc.
851 New Holland Avenue
Box 3535
Lancaster, Pennsylvania 17604 U.S.A.

*Distributed in the Rest of the World by*
Technomic Publishing AG

Printed in the United States of America
10  9  8  7  6  5  4  3  2  1

Main entry under title:
  Composite Panels/Plates: Analysis and Design

A Technomic Publishing Company book
Bibliography: p.
Includes index p. 353

Library of Congress Card No. 86-50497
ISBN No. 87762-476-3

# TABLE OF CONTENTS

v

# PREFACE

This book presents principles and methods for the analysis and design of composite construction and shows how these theories may be used in practical design problems.

Construction with composite materials has been extensively used in the aerospace industry, and its application in building is rapidly increasing. This great interest in composite materials is explained by the following quote from Dietz [Composite Engineering Laminates, MIT Press]: "Demands on materials imposed by today's advanced technologies have become so diverse and severe that they often cannot be met by simple single-component materials acting alone. It is frequently necessary to combine several materials into a composite to which each constituent not only contributes its share, but whose combined action transcends the sum of the individual properties, and provides new performance unattainable by the constituents acting alone. Space vehicles, heat shields, rocket propellants, deep submergence vessels, buildings, vehicles for water and land transport, aircraft, pressure tanks, and many others impose requirements that are best met, and in many instances met only by composite materials."

The concept of composite structures is not new. A piece of laminated wood stored at the Metropolitan Museum of Art, New York, was found at Thebes and shown to belong to the Eighteenth Dynasty (about 1500 B.C.). On its bottom are five pieces glued and laminated to a heavier piece, which is made of glued veneer. Another example is the ancient sarcophagus of wood veneers laminated to a heavier wood substrate and now in the Boston Museum of Fine Arts. The subtle principles of a composite system were understood intuitively and empirically and are now the subject of major research. Sandwich construction is also characterized by the use of two layers of strong material, called "faces," between which a thick layer of a light-weight and comparatively weak core is sandwiched. Such composite construction has found many applications in the aerospace and building industries. It is used in the construction of helicopter rotor blades, helicopter flooring, aircraft wings, deck panels, fire walls, access doors, and baggage racks. Movable objects, like tank trailers that transport bulk milk and fruit juices, ladders for use in telephone line service, and truck trailer panels and doors, are being built using composite sandwich panels. Architects are recognizing the fact that the composite concept is well suited for curtain wall applications. In addition, it is employed widely in structural walls, roofs, and floorings for house trailers, small boat hulls, shipboard doors and bulkheads, table tops, and furniture. Folding plate roofs can also be built using composite panels. One-story homes, motels and similar buildings have been successfully built with composite panels. As an example, a motel and a restaurant were fabricated in

Canada and shipped for erection in the Caribbean Islands. Another example is the widespread use of composite panels in the southern United States, and in northern Canada, and in Europe. In multi-story buildings, composite panels have a place in the total system up to perhaps four floors, but only two-story buildings have been built.

Answers to many problems related to structural systems may be found in composite construction, which offers several advantages. Virtues related to environmental requirements may include structural integrity, durable finishes, weather-tightness, dimensional stability, sound or microwave absorption. From the point of view of structural performance, composite construction is an efficient structural design due to the great stiffness achieved by different geometrical arrangements of the most highly stressed elements. Other advantages of composite construction may include high strength-to-weight ratios, increased fatigue life, endurance, low moisture permeability, electrical insulation, color processability, reduced shop labor, simplified materials handling, and potentially lower costs to the house manufacturer and for transportation.

Although many types of composite structures have been invented, developed, tested, and marketed for use in the construction trade, there exist no design codes or specifications to assist designers. Thus, a designer must determine on his own to criteria for selection and design which are most appropriate for his particular application. Normally, he may gain valuable assistance from the wealth of information accumulated since the beginning of the Second World War. However, in the final analysis he makes his own determination of the suitability of a specific design for a particular application under investigation, together with, in many cases, a prototype test program to verify the adequacy of the design developed. For composite construction, because of the wide variety of available options for configuration, types of materials, and different fabrication processes, it is vital that the designer carefully establish his desired functional and performance requirements as accurately and in as much quantified detail as possible. A substantial part of the process involves the determination of the structural response and the proportioning of the elements which make up the component.

This book is aimed at assisting the designer to fully understand the different structural aspects of composites, to select suitable materials and structural design criteria, and to analyze the effects of loads and restraints. The book presents the state-of-the-art in this field. Its objective is to provide simplified formulas and design aids related to composite structures. Mathematical derivations are in a form suitable for this objective. Methods of analysis and basic assumptions underlying each theory are discussed. Particular attention has also been paid to provide the complex mathematical results in a practical, simple, exact form suitable for design offices. Parameters have been chosen in the formulas in such a way as to arrive at convenient forms, tables, and graphs. Extensive references have been presented at the end of each chapter. The results of the author's experience during the past decade have been included. Because the structural applications of composites are numerous, the implementation of all theories is devoted to laminated composites, which covers a broad spectrum in this field. It is felt that the book can easily be followed by persons not already familiar with its subject, including senior under-graduate, graduate students, and practicing engineers. A background in strength of materials and the fundamentals of structural analysis is sufficient to comprehend the theories presented.

This book is a result of the excitement and challenge that the author experienced in working in the field of composites. I would like to extend very special thanks to Professor Paul Cheremisinoff, from the New Jersey Institute of Technology for his continuous en-couragement throughout all aspects of writing. My sincere gratitude goes also to Mrs. Charmine S. for her outstanding job in typing the manuscript.

Finally, truly unbounded thanks are due to my wife and son for their love and patience over the years, which made the writing of this book a reality.

RAFAAT M. HUSSEIN

# Composites in Construction

## 1.1  INTRODUCTION

There is growing interest in the development and applications of composite construction. Composite action, for example, is based on the concept of combining dissimilar materials to form a composite fulfilling specific design requirements.

A three-layer laminated panel, as a special form of composite actions, is characterized by the use of two thin layers of strong material, denoted as faces, between which a thick layer of lightwieght and comparatively weak core is sandwiched as shown in Fig. (1). In a structural panel of this type, the faces resist bending moments and in plane compressive or shear forces. The shear forces normal to the plane of the panel are resisted by the core, which also stabilizes the faces against buckling (Fig. (2)). This type of construction is efficient structurally due to the large stiffness achieved by spacing apart the most highly stressed elements, namely, the faces. The basic principle is much the same as that of an I-beam.

## 1.2  COMPOSITE LAMINATED PANELS IN CONSTRUCTION

Answers to many problems related to construction may be found in composite laminated panels which offer several advantages. Panel virtues related to environmental requirements may include structural integrity, durable finishes, weather-tightness, dimensional stability, sound or microwave absorption. From the point of view of structural performance, a composite laminated construction is an efficient structural design due to the large stiffness achieved. Other advantages may include high strength-to-weight ratios, increased fatigue life, endurance, low moisture permeability, electrical insulation, color processability, reduced shop labor, and potentially lower costs to the house manufacture and transportation.

As with any other component, there are some factors to be considered. When the core or facing materials may lead to corrosion problems, special pretreatments are required. No foam is fireproof, but many of them can be made nonflammable. Polyester laminated panels were tested, both painted and unpainted, by the Forest Products Laboratory, and it was found that the unpainted panels deteriorated the most in three year's weathering; the edgewise compressive strength reduced by 40 percent

**1**

and the flexural strength by 30 percent. Painting the other polyester panels reduced the loss in edgewise compressive strength to 22 percent, but the reduction in flexural strength was still 30 percent. Similar effects on panel strengths were observed due to the effect of moisture contents. Although several advantages can be attributed to most plastic foams, their resistance to chemical agents should be considered in the manufacturing process.

## 1.3   MATERIAL REQUIREMENTS

A laminated composite panel in building systems should combine the features stated before. Constituents contribute to achieve these features as well as fulfilling the basic principle of this type of construction. A list of the required properties of each layer in a laminated composite panel is presented in Table 1.

Table 1

REQUIRED PROPERTIES

| Panel | Skin | Core | Bonding | Frame | Connections |
|---|---|---|---|---|---|
| ·Strength | ·Strength | ·Strength | ·Strength | ·Strength | ·Cheap |
| ·Rigidity | ·Resistance | ·Rigidity | ·High peel | ·Rigidity | ·Simple |
| ·Repair- | to local | ·Imperme- | ·Strength | ·Insulation | ·Structural |
| ability | damage | ability | ·Durability | ·Repair- | ·Weather- |
| ·Durability | ·Weather | ·Nonrotting | ·Low creep | ability | tight |
| ·No loss of | resistance | ·Fire res- | ·No loss of | ·Weather | |
| strength | ·Nonrotting | istance | strength | resistance | |
| up to high | ·Fire res- | ·Sound Ab- | up to high | ·Sound | |
| tempera- | istance | sorption | temperature | absorption | |
| ture | ·Imperme- | ·Insulation | ·Fatigue | | |
| ·Minimum | ability | ·Durability | ·Long life | | |
| weight | ·Durability | ·Minimum | ·Suitable for | | |
| ·Fatigue | ·Reflective | weight | varied | | |
| ·Vibration | ·Sound ab- | ·Incombust- | materials | | |
| | sorbing | ible | ·Adhesion main- | | |
| | | ·Can accept | tenance during | | |
| | | adhesion | contact with | | |
| | | | moisture or | | |
| | | | water | | |
| | | | ·Chemical sta- | | |
| | | | bility regard- | | |
| | | | ing both faces | | |
| | | | and core | | |

## 1.4   PANEL MATERIALS

A laminated composite panel development provides the opportunity of not favoring any one material, but rather employing most of the available materials. Therefore, a desired design requirement can be achieved by more than alternatives of material combinations.

Some of the materials that have been or might be used are listed in Table 2.

Table 2

MATERIALS FOR LAMINATED PANELS

| Faces | Core | Bonding | Frame | Connections |
|-------|------|---------|-------|-------------|
| ·Plywood | ·Kraft paper | ·Latex | ·Timber | ·Nails |
| ·Gypsum | honeycomb | ·Neoprene base | ·Steel | ·Screws |
| board | ·Balsa wood | contact cement | ·Aluminum | ·Bolts |
| ·High pres- | ·Light weight | ·Epoxy resins | | ·Locking |
| sure lami- | concrete | | | devices |
| nates | | | | ·Cover strip |
| ·Painted | | | | |
| steel sheet | | | | |
| ·Stainless | | | | |
| steel sheet | | | | |
| ·Asbestos | | | | |

## 1.5   POLYMERS  IN  COMPOSITE  PANELS

Combining several materials into one composite unit is not a new idea.   The principles of a laminated system were understood intuitively and by experience by ancients.   Spaced fibers were developed by a Frenchman named Dulcan in 1820. During World War II, demands on efficient use of labor and materials imposed by the aircraft industry resulted in promoting the use of laminted panels with plywood faces and honeycomb cores.   However, by today's advanced technologies, new materials, most of which are polymers, were developed and this resulted in an impact on the area of laminated construction.

Laminated panels may be made entirely of plastics or a combination of plastic with other materials.   For example, the core may be made of rigid foams and skins of aluminum. A list of other plastics for use in laminated panels is given in Table 3.

## Table 3

### POLYMERS FOR LAMINATED PANELS

| Faces | Core | Bonding | |
|-------|------|---------|--|
| ·Phenolic-Asbestos Laminates | ·Fibre Glass | ·Latex | ·Phenlic-neoprene |
| ·Epoxy-Glass Cloth Laminates | ·Expanded Polystyrene | ·Neoprene base contact cement | ·Phenolic-vinyl |
| ·Polyester-Glass Fiber Laminates | ·Expanded Polyurethane | ·Alkyd | ·Polyacryl-onitrile |
| ·PVC | ·Expanded PVC | ·Acrylate | ·Polamide |
| ·PMMA | ·Acrylonitrile styrene | ·Caselin | ·Polyethy-lene |
| ·Laminated wood-plastic | ·Cellulose Acetate | ·Cellulose-nitrate | ·Polyimide |
| ·FRP | ·Epoxy | ·Cellulose-vinyl | ·Polyvinyl-acetate |
| ·Polyamide (nylon) | ·Methylmethacrylate styrene | ·Epoxy | ·Polyvinyl-butyral |
| ·Plastic Clad Plywood | ·Phenolic | ·Epoxy-novalac | ·Resorcinol-phenol formalde-hyde |
| | ·Polyethylene | ·Epoxy-phenolic | |
| | ·Polypropylene | ·Epoxy-polyamide | ·Silicone |
| | ·Polystyrene | ·Epoxy-polysulfide | ·Vinyl buy-ralphenolic |
| | ·Polyvinyl Chloride | ·Epoxy-silicone | ·Vinyl Copo-lymers |
| | ·Silicone | ·Melamine-formaldehyde | ·Urea formalde-hyde |
| | ·Urea Formaldehyde | ·Phenolic | ·Urethane |
| | ·Urethane | ·Phenolic-butadience acrylonitrile | |

The behavior of a laminated panel depends on its constitutents, its geometry, and the applied technology in the manufacturing process. The virtues of some polymers for the use in such panels will be discussed next.

## 1.5.1   SURFACE FIBERS

Among the synthetic materials often used for the outer fibers of composite panels are:

Poly (vinyl chloride) - PVC - and vinyl chloride copolymers
Poly (methyl methacrylate) - PMMA
Laminated wood-plastic
Fibre-reinforced plastics - FRP

POLY (Vinyl Chloride):

Unplasticized PVC is a hard horny material, insoluble in most solvents and not easily softened. If it is mixed while hot with certain plasticizers, it forms a rubbery material.

PVC has a tendency to liberate hydrochloric acid, particularly at temperatures approaching the upper limit of serviceability. To prevent this reaction stabilization with acid neutralizers or absorbers is necessary. PVC is nonflammable, moistureresistant, odorless, tasteless, and offers exceptional resistance against a great number of corrosive media.

Advances in rigid PVC formulation and in processing equipment design have made available a new dimension in building panels to the building industry. PVC elements lend themselves admirably to use for roofing, siding, various shelter and enclosure structures, and for interior panelling; in fact, they have already caused some revolutionary changes in building concepts.

Rigid PVC panels have some outstanding advantages, as follows: attractive appearance, weatherability, lightweight, corrosion and fire resistance, easy to clean and easy to install. New rigid PVC-resistant to higher temperature - contains clorinated PVC; these plastics may be used at temperatures up to 100°C.

Copolymerization of vinyl chloride with other vinyl monomers leads to some products with improved properties.

POLY (Methyl Methacrylate):

This is a clear, colorless, transparent plastic with a high softening point, good impact strength and weatherability. Sheets of PMMA are commonly made by extrusion.

The resistance of PMMA to outdoor exposure is outstanding and it is markedly superior to other thermoplastics. It has a low water absorption and the abrasion resistance is roughly comparable with that of aluminum. Like other polymers, PMMA is thermally insulating.

In sheet form, PMMA and other acrylic polymers can be softened at high temperatures and formed (usually vacuum formed) into practically any desired three-dimensional shape.

These polymers have excellent transparency (total luminous transmission up to 93%), while polystyrenes in their initial stage rarely exceed 88% - 90%. Due to the latter characteristics, PMMA have been used for many lighting and glazing applications.

PLASTIC CLAD PLYWOOD:

The properties of this well-known and proved product are essentially those of solid wood except that the cross-piles greatly increase strength and stability in the cross-grain direction at the expense of some reduction in the grain direction.

Plastic clad plywood touches almost all of us. This laminate is the familiar work surface of the kitchen and many restaurant tables. Again, the advantages of the plywood and light weight are enhanced for the specific use, by the hardness, decorative quality and wear resistance of the plastic outer layer.

COMPOSITES:

Composite materials are being applied in increasing amounts for essentially the same reasons as for other applications; namely, they provide properties and behavior not attainable in single-phase materials, or they provide these features more efficiently, or at lower cost, or both.

The type of composite materials or composite structures principally found in building are:
Fibrous: Fibres embedded in a continuous matrix laminar;
Layers of materials blended together and possible interpenetrated by a building material particulate;
Particles embedded in a continuous matrix.

Composites have three outstanding advantages: ease of fabrication, high fracture energy, and potential low cost. The latter is particularly true for glass-reinforced resins, glass being by far the most common fibre used in resin-matrix composites. Added advantages of the resin-matrix composites often include the low density, low electrical and thermal conductivity, transluscence aesthetic, color effects, and corrosion resistance. The resin-matrix composites are highly formable, and their fracture energy is enormous. Table 4 illustrates the unusual fracture energy possibilities of resin-matrix composites.

Mechanical properties of a composite depend primarily on the type of fibres, their quantity in the laminate and their direction. The type of fibre, that is, its tension strength and its modulus, determine these properties in the finished laminate and the percentage value of these fibres is directly proportional to the tensile strength and to the modulus. These are short-fibre-reinforcements such as wollastonite, asbestos, or continuous filament reinforcements such as glassfibre, basalt fibres, high modulus organic fibres, aluminum oxide and other ceramic filaments, metal filaments.

The more important fibrous materials for advanced composites are compared in Table 5.

Table 4

FRACTURE ENERGY

| Material | Energy to Propagate a Crack $J/cm^2$ |
|---|---|
| Glass | 0.061 |
| Epoxy resin | 0.61 |
| Metal | 183.00 |
| Glass-reinforced epoxy | |
|     Parallel fibre | 3.1 |
|     Perpendicular | 430.00 |

Table 5

PROPERTIES OF THE MORE COMMON FIBRES USED IN REINFORCED PLASTICS

| Material | Tensil Strength $(N/cm^2 \times 10^3)$ | Modulus $(N/cm^2 \times 10^6)$ | Density $(g/cm^3)$ |
|---|---|---|---|
| E glass | 345 | 7.2 | 2.55 |
| S glass | 450 | 8.6 | 2.5 |
| PRD-49-III | 275 | 13 | 1.45 |
| Boron | 275-310 | 38-41 | 2.4 |
| Carbon | 103-310 | 69-62 | 1.4-1.9 |
| Steel Wire | 206-512 | 20 | 7.7-7.8 |

There are several glass formulations that have been used as fibres in polymeric composites, these include A, C, D, E, M, and S glass. In terms of advanced composites, E and S glass are not the most important. E glass, of low alkali oxide content (composition given in Table 6), has become the most widely used formulation for both textile and industrial applications.

Between fibre and matrix (resin), there is an interfacial bond (a coupling agent) whose function seems to prevent the complete adhesion of the fibre to resin so that relative motion is possible.

The cracks, instead of penetrating through the matrix and directly through the fibres, are deflected along the length of the fibres.

Table 6

FORMULATION OF E GLASS

| Constituent | Content (%) |
|---|---|
| Silicon dioxide | 52-56 |
| Calcium oxide | 16-25 |
| Aluminum oxide | 12-16 |
| Boron oxide | 8-13 |
| Sodium and potassium oxides | 0-1 |
| Magnesium oxide | 0-6 |

The most common coupling agents for glass are the organosilanes and the chrome complexes; the last are cheaper and preferred in a great number of applications.

The availability of a wide variety of manufacturing techniques has undoubtedly been an important factor in the steady growth of the reinforced plastics industry, and it is very likely that it will exert considerable influence on the progress of advanced fibrous composites and laminated panels.

Table 7 shows the relative efficiencies of various forms of fibrous reinforcements: the superiority demonstrated by the continuous-oriented fibre structures is apparent.

It has already been pointed out that the properties of a fibrous composite depend upon the characteristics of fibres, matrix and coupling agent. Several polymeric matrices are used and may be divided into: thermosets and thermoplastics. Sometimes, polyblends are used.

Cross-linked polyesters, phenol-formaldehyde, melamine-formaldehyde resins, epoxide and silicone polymers are among the most common in the thermosetting field.

The characteristics of these polymers may be seen in Table 8.

The second group, the thermoplastics, are poor in one or more of the following engineering properties: creep resistance, rigidity and tensile strength, dimensional stability, impact strength, maximum service temperatures and hardness. The use of reinforcement is one way of overcoming some of these difficulties. Table 9 shows some of the relative strength to weight ratios of various materials and illustrates well the advantages to be gained by using thermoplastics in composite form. Polyamide nylon 66 is a good example of a thermoplastic amendable to modification. A lot of properties can be improved by incorporating 20% - 40% glassfibre.

In many cases, the new composites have to demonstrate their superiority over timber and its derived products, such as hardboard, chipboard and plywood. Polymer composites have the advantages that they are generally immediately ready for use and

do not require finishing processes. Moreover, they usually need very little maintenance after exposure to weather.

Table 7

TYPICAL COMPOSITE EFFICIENCIES ATTAINED IN REINFORCED PLASTICS

| Fibre Configuration | Fibre Length | Total Fibre Content (by volume) $V_f$ | $F_{long}$ ksi $(N/cm^2 \times 10^3)$ $F_{theor}$ | $F_{test}$ | Composite Efficiency (%) |
|---|---|---|---|---|---|
| Filament-wound (un-directional) | Continuous | 0.77 | 310(214) | 180(124) | 58.0 |
| Cross laminated fibres | Continuous | 0.48 | 197(136) | 72.5(50.0) | 36.0 |
| Cloth laminated fibres | Continuous | 0.48 | 197(136) | 43.0(29.6) | 21.8 |
| Mat laminated fibres | Continuous | 0.48 | 197(136) | 57.2(39.4) | 29.0 |

*(continued)*

| | | | | | |
|---|---|---|---|---|---|
| Chopped fibre systems (random) | Non-continuous | 0.13 | 60.7(41.8) | 15.0(10.3) | 24.7 |
| Glass Flake composites | Non-continuous | 0.70 | 165.5(114.1) | 20.0(13.8) | 12.1 |

$F_{long}$ = Ultimate tensile strength in direction of greatest fibre content (longitudinal), if there is one.

Theoretical strength based on 'Rule of Mixtures':

$F_{theor} = V_f S_f + (1 - V_f)S_m$

where: $S_f$ 400.(275.8)ksi($N/cm^2$ x $10^3$) - typical boron or carbon fibre strength

$S_m$10.(6.9)ksi($N/cm^2$ x $10^3$) - typical resin strength

$F_{test}$ = typical experimental strength values

Composite efficiency = $(F_{test}/F_{theor})$ x 100

Table 8

## CHARACTERISTICS AND USES OF REINFORCED THERMOSETTING RESINS

| Resins | Characteristics | Uses | Limitations |
|---|---|---|---|
| Diallyl phthalate polymer | Good electrical properties; dimensional stability; chemical and heat resistance | Prepegs, ducting, radomes, aircraft, missiles | |
| Epoxy | Good electrical properties; chemical resistance; high strength | Printed circuitboard tooling filament winding | Require heat curing for maximum performance |
| Melamine formalde-hyde | Good electrical properties; chemical and heat resistance | Decorative, electrical (arc and track resistance), circuit breakers | |
| Phenolic | Low cost; chemical resistance; good electrical | General-diverse | Dissolve in caustic unless |

| | | |
|---|---|---|
| properties; heat resistance; nonflammable; can be used to 350-400°F (177-204°C) | mechanical and electrical applications | specially treated |
| Polyester | Good all-round properties; ease of fabrication; low cost; versatile | Corrugated sheeting, seating, boats automotive, tanks, and piping, aircraft, tote boxes, | Degraded by strong oxidizers, aromatic solvents concentrated caustic |
| Silicone (qv) | Heat resistance; good electrical properties | Electrical, aerospace | |

Table 9

RELATIVE STRENGTH-TO-WEIGHT RATIOS OF VARIOUS TYPES OF MATERIALS

| Material | Strength-to-Weight Ratio Relative to Polycarbonate |
|---|---|
| Polycarbonate | 1.00 |
| Polystyrene | 1.09 |
| Nylon | 1.24 |
| Styrene-acrylonitrile copolymers | 1.48 |
| Brass (yellow cast) | 1.52 |
| Zinc alloys (cast) | 1.67 |
| Glass/polystyrene | 1.71 |
| Glass/polycarbonate | 1.76 |
| Glass/styrene-acrylonitrile copolymers | 1.95 |
| Magnesium | 2.19 |
| Aluminum | 2.52 |
| Glass/nylon | 2.62 |

## 1.5.2 INNER FIBERS

Different materials are used to form inner layers in composite panels. Foamed or cellular plastics, honeycomb and others.

Cellular structural materials are produced by making 'solid' foams from various polymers such as: phenolic resins, cellulose-acetate, polystyrene, poly (vinyl chloride), polyurethane and others. Depending on manufacturing methods and composition, these structures may have different properties, such as:
1. Relative weight
2. Flexibility or rigidity
3. Preparation from thermosetting or thermoplastic polymer
4. Presence or absence of plasticizers
5. Method of cells formation
6. Nature of cells in the solid foam
7. Whether foamed in mould in place or freely-expanded

Polystyrene and polyurethane are used to make a wide variety of foamed products. Most polystyrene foams are based on foamed in-place beads, made for suspension polymerization in the presence of a foaming agent. Subsequent heating softens the resin and volatizes the foaming agent.

Polyurethane foams are expanded by a somewhat different chemical reaction. These expanded plastics may have separate interconnected or partially interconnected cells. Rigid polyurethane foams are resistant to compression and may be used to reinforce hollow structural units with a minimum of weight. In addition, they consist of closed cells and have low rates of heat transmission. They develop excellent adhesion when they are formed in voids or between sheets of material. Finally, they are resistant to oils and do not absorb appreciable amounts of water. These properties make the rigid foams valuable for prefabricated laminated structures used in the building industry for thermal insulation.

Polymers were used in an experimental study of laminated foam coextrusion using a sheet-forming die with a feedback. The polymers used in the experiment were low density polyethylene for the outer layers and ethylene-vinyl-acetate copolymer for the foamed core component. The study has shown that cell size and its distribution in the foamed core and the mechanical properties of the laminated foam product can be controlled by a judicious choice of the thickness ratio of the core to the skin components, the melextrusion temperature, and the concentration of chemical blowing agent.

The production of panels by foaming placed between two skins, takes advantage of the capability of polyurethanes to enter the moulds as a liquid or semi-expanded froth and subsequently expand to fill every section of the mould cavity. Obviously this provides a degree of freedom in panel design which is not available with lamination from slab stock. Complex-shaped panels which would require an impractical amount of cutting and with slab are produced rather simply by foaming in place.

Honeycomb consists of thin foils in the form of hexagonal cells perpendicular to the faces. Treatment of the draft paper with phenolic resin assures adequate durability if the panels are made and assembled in place to remain dry, or at least drain freely at edges and joints. Adhesives should be chosen to 'wet' the honeycomb and adjacent layer, in order to form a fillet to distribute the shear stresses on the thin paper edges. Metals can be fully stabilized by the high strength of these hexagonal paper structures but the open cells provide little bending resistance. For many uses paper-phenolic honeycomb remains the standard for structural purpose.

## 1.5.3 BONDING

These are the bonding elements between layers of composite structure. The bond allows two thin faces to be used structurally together with the core which makes the panel tough.

There are a lot of synthetic polymers which may be used to manufacture laminated panels. They may be thermosets or themoplastics as follows:
a) Urea-formaldehyde polymers: they are the cheapest but have inferior properties because of their sensitivity to moisture.
b) Phenol-formaldehyde resins and their derivatives: they are cheap, but need heating or acid catalysts, which may attack outer and/or inner elements.

c) Resorcinol-formaldehyde resins: they may be used at room temperature with neutral catalysts which do not influence the panel elements.
d) Unsaturated polyesters: they form the basis of glass-reinforced plastics used for constructional purposes, but are also marketed as adhesives.
e) Epoxy polymers: They have a good adhesion with metals and good mechanical resistance, but are expensive. They also have the same draw-backs with the unsaturated polyesters; that is, to be made of two different components.
f) Acrylic copolymers: they may be used as adhesives which cure through solvent evaporation or mixture or monomers.
g) Polychloroprene: it is the most used elastomer-adhesive; it is cheap and has the advantage that there is no time restriction to be assemblage.
h) Polyblends, like mixture of epoxy polymers and elastomers (polychloroprenes), they appear good as resistant to water, but are expensive. Their use eliminates the necessity of pressing after gluing.

Adhesives for composite panels are prepared as solutions, lattices or films. With the last one a constant thickness will be needed on the glued surface to ensure that the strength is uniform. However, films are more expensive than liquid adhesives and are useful only for honeycombs.

Recommended curing conditions will vary depending on the used materials. In many cases, modification of a given curing procedure can be brought about if advantageous to the producer. The adhesive manufacturer should be consulted for specific recommendations on his adhesive system and the recommendations should be followed exactly.

## 2

# Simply Supported Composite Plates

## 2.1  INTRODUCTION

In this chapter the governing equations of composite plates are presented. The developments are based on energy methods and plates theory [29]. Navier's solutions [28, 29] for simply supported composite plates are obtained for the following loading types: (i) uniform distributed load; (ii) concentrated load; (iii) hydrostatic pressure; (iv) partial load; and (v) strip load. For the last three loading types, solutions have not been found in the literature.

Existing techniques for the analysis of composite plates include: (i) variational method; (ii) the partial deflection theory, and (iii) numerical methods such as the finite element and the finite difference methods. A survey for each of the three methods is given next.

First, Reissner [23, 24, 25] by using the variational theorems, studied the effect of transverse shear deformation on the bending of elastic plates. Taking into consideration the deformability of such plates due to transverse shear, a system of differential equations for deformations were obtained based on Castigliano's theorem of least work combined with the Lagrangian multiplier method of the calculus of variations. The principal contribution of Reissner, in addition to the equations obtained, was to furnish three boundary conditions to be satisfied by a solution at each edge of the plate instead of two as Kirchoff established. By similar variational theorems other analyses were conducted. For instance, Chong [5] in his approach based on the principle of minimum complementary energy combined with the Lagrangian multiplier method achieved similar results to Reissner's. Furthermore, a governing linear differential equation of the sixth-order for the deflection was derived, from which a solution can be obtained through the use of techniques developed for classical plates [28, 29]. Another example of solving a plate by applying the variational theorems, is the analysis carried out by Ueng [30].

Second, since the development of the partial deflection theory by March [15], its application was adopted extensively by many others [3, 7, 21]. According to this theory, any line that is initially straight and normal to the middle surface will remain straight after the deformation of the plate but not necessarily normal to the deformed middle surface. Consequently, the plate deflection can be considered to be composed of two components: bending and shear deflections. The former corresponds to the deflection of a composite plate rigid in shear while the latter to that of a plate infinitely stiff in bending. Two distinct methods can be used to determine the partial deflections: (i) Plantema's approach [21] where a governing differential equation was developed for each component of the deflection, and (ii) Allen's approach [3] where the shear strain is related to the total slope by unknown parameters which can be determined by using the principle of minimum total potential energy.

Finally, to avoid the complex mathematical operation of the two previous methods, numerical techniques have been widely utilized. Among them, the finite element method has enjoyed wide acceptance. Both the assumed displacement and hybrid approaches have been applied in deriving the stiffness matrix of finite elements of plates [1, 4, 6, 9, 10, 12, 13, 14, 16, 17, 18, 19, 27]. Several finite element computer programs to solve composite plates have been written [1, 4, 9, 18]. The other popular numerical technique is the finite difference method [2, 28]. This method has a long tradition and is a well-known technique for solving boundary value problems.

In summary, theories of bending of composite plates are well established by a number of authors. However, these theories show differences with regard to the mehtod of analysis, the equations developed, and the solution methods.

## 2.2  ASSUMPTIONS

Composite plates under transverse load present a problem in three dimensional elasticity. However, by introducing assumptions regarding the stress or displacement distribution across the plate thickness, the problem can be reduced to a two dimensional one. In this book, the mathematical analysis of three layered composite plates is presented. In such laminar construction, a combination of alternating dissimilar composite materials is assembled and intimately fixed in relation to each other so as to use the properties of each to obtain specific structural advantages for the whole system.

The relevant assumptions forming the basis for deriving the governing equations which are solved later on in the chapter are as follows:
(1) The outer laminae are thin in comparison with the overall depth. This implies that the flexural rigidity of each lamina about its own middle surface is negligible, and consequently, the inplane stresses resisted by each of them are uniformly distributed across its thickness.
(2) The inplane stresses in the middle lamina (called core) are negligible. This means that the core material is soft relative to the outer material. Thus, the transverse shear stresses are constant across the thickness. And since the stress normal to the plate surface has an insignificant effect on the bending stresses [26], it follows that the normal to the middle plane of the plate remains straight and inextensible after deformations, as was proposed by March [15]. The general solutions developed in this chapter are applicable to composite plates whose inner

laminae experience warping. A modified shear modulus must be used in this

particular case, and can be calculated, however, from an expression developed by Allen [3].

(3) The composite plate is assembled such that full interaction at the interfaces is provided through a rigid adhesive. The effects of interlayer deformations on the response of composite plates will be studied in chapter IV.

(4) The two outer laminae are of equal thickness and made of the same material.

(5) Materials are homogeneous and linearly elastic.

(6) Deformations are small.

## 2.3 GOVERNING DIFFERENTIAL EQUATIONS FOR DEFLECTION AND SHEARS

The moment expressions, equilibrium and governing equations for orthotropic composite plates with dissimilar outer laminae are well established [3, 5, 21, 23, 30]. The mathematical treatments presented are complex for an average practicing engineer, and therefore, these equations are derived in this section by following a simple approach, and then reduced to represent an isotropic composite plate for which the aforementioned assumptions are applicable.

Consider now a simply supported laminar plate as shown in Fig. 2.1. The non-zero stress components in the outer laminae are $\sigma_x$, $\sigma_y$, $\tau_{xy}$ and in the core $\tau_{xz}$, $\tau_{yz}$. The strain energy in the plate, considering small deformations and rigid bondings, is equal to the strain energy in the three laminae [3, 5, 9, 23, 30].

$$U = \frac{1}{2} \int_f \sum_{i=1}^{2} [t_i (\frac{\sigma_{xi}^2}{E_{xi}} + \frac{\sigma_{yi}^2}{E_{yi}} - \frac{2\nu_{xyi}}{E_{xi}} \sigma_{xi} \sigma_{yi} + \frac{\tau_{xyi}^2}{G_i})] \, dA +$$

$$\frac{1}{2} \int_c (\frac{\tau_{xz}^2}{G_{xz}} + \frac{\tau_{yz}^2}{G_{yz}}) \, dv$$

in which

    f and c    denote the facing and core, respectively;

    $\nu_{xy}$ = Poisson's ratio of outer material;

    E = Young modulus of outer material;

    G = shear modulus;

    t = thickness of outer lamina;

    $\sigma$ = normal stress;

    $\tau$ = shear stress;

1, 2      = subscripts denote faces 1 and 2, respectively (Fig. 2.1).

The resultant moments per unit length can be expressed in terms of the uniform stresses across the thickness t, as:

$M_x$          =                                                $t_1 h\sigma_{x1} = -t_2 h\sigma_{x2}$

$M_y$          =                                                $t_1 h\sigma_{y1} = -t_2 h\sigma_{y2}$

$M_{xy}$       = $-t_1 h\tau_{xy1} = t_2 h\tau_{xy2}$

in which $h = t_c + \frac{1}{2}(t_1 + t_2)$, as shown in Fig. 2.2(a).  Shearing forces per unit length can also be expressed in terms of transverse shearing stresses as [3]:

$Q_x$          = $h\tau_{xz} = G_{xz}h\gamma_x' = G_{xz}(h^2/t_c)\gamma_x$

$Q_y$          = $h\tau_{yz} = G_{yz}h\gamma_y' = G_{yz}(h^2/t_c)\gamma_y$

in which

$\gamma_x', \gamma_y'$      = shearing strains;

$\gamma_x, \gamma_y$      = slope of the shearing deflection surface, as shown in Fig. 2.2(b).

Substituting the resultant forces-stresses relations in the strain energy equation and simplifying its terms yields:

$$U = \int_A [(M_x^2/D_x') + (M_y^2/D_y') - 2(M_x M_y/D_\upsilon) + (M_{xy}^2/D_{xy}) + (Q_x^2/S_x) + (Q_y^2/S_y)] \, dA$$

in which

$D_x'$      = $h^2[(1/E_{x1}t_1) + (1/E_{x2}t_2)]^{-1}$

$D_y'$      = $h^2[(1/E_{y1}t_1) + (1/E_{y2}t_2)]^{-1}$

$D_{xy}$   = $h^2[(1/G_{xy1}t_1) + (1/G_{xy2}t_2)]^{-1}$

$D_\upsilon$      = $h^2[(\upsilon_{xy1}/E_{x1}t_1) + (\upsilon_{xy2}/E_{x2}t_2)]^{-1}$

$S_x$      = $G_{xz}h^2/t_c$

$S_y$      = $G_{yz}h^2/t_c$

According to fifth assumption in section 2.2, the composite  constituents are linearly elastic.  Thus from Castigliano's second theorem it follows that:

$\partial U/\partial M_x = -\partial\theta_x/\partial x$

$$\partial U / \partial M_y = -\partial \theta_y / \partial x$$

in which

$$\theta_x = (\partial w / \partial x) - \gamma_x$$
$$\theta_y = (\partial w / \partial y) - \gamma_y$$

hence

$$(M_x / D_x') - (M_y / D_\upsilon) = -\partial \theta_x / \partial x$$
$$(M_y / D_y') - (M_x / D_\upsilon) = -\partial \theta_y / \partial y$$

Solving these equations in $M_x$ and $M_y$ yields

$$M_x = -D_x \left( \frac{\partial \theta_x}{\partial x} + \frac{D_y'}{D_\upsilon} \frac{\partial \theta_y}{\partial y} \right) \qquad (2.1)$$

$$M_y = -D_y \left( \frac{\partial \theta_y}{\partial y} + \frac{D_x'}{D_\upsilon} \frac{\partial \theta_x}{\partial x} \right) \qquad (2.2)$$

in which

$$D_x = \left( \frac{1}{D_x'} - \frac{D_y'}{D_\upsilon^2} \right)^{-1}$$

$$D_y = \left( \frac{1}{D_y'} - \frac{D_x'}{D_\upsilon^2} \right)^{-1}$$

In addition, the twisting moment can be obtained using the same theory as follows:

$$\frac{\partial U}{\partial M_{xy}} = \frac{\partial \theta_x}{\partial y} + \frac{\partial \theta_y}{\partial x}$$

from which

$$M_{xy} = D_{xy} \left( \frac{\partial \theta_x}{\partial y} + \frac{\partial \theta_y}{\partial x} \right) \tag{2.3}$$

Expressions for the shearing forces are given before. The same results can be derived using Castigliano's theorem. In the moment and shear forces expressions, $\theta_x$ and $\theta_y$ represent the normal slopes of the plate due to moment effects. $D_x$ and $D_y$ may be regarded as flexural rigidities of the plate, and $D_{xy}$ as its twisting stiffness. The $S_x$ and $S_y$ are the shear rigidities. In the particular case of isotropic composite plates with similar facings:

$$E_f = E_{x1} = E_{x2} = E_{y1} = E_{y2}$$

$$\upsilon = \upsilon_{xy1} = \upsilon_{xy2}$$

$$t_f = t_1 = t_2$$

$$h = t_c + t_f$$

$$G_f = G_{xy1} = G_{xy2} = E_f/2(1 + \upsilon)$$

$$G_c = G_{xz} = G_{yz}$$

Consequently, the stiffness constants become:

$$D = D_x = D_y$$

$$= E_f t_f h^2/2(1 - \upsilon^2)$$

$$D_{xy} = (0.50)D(1 - \upsilon)$$

$$S = S_x = S_y$$

$$= G_c(h^2/t_c)$$

The equilibrium conditions of the laminar element in Fig. 2.2(a) are the same as those for homogeneous plates. Consider, for example, the equilibrium of moments in x-direction

$$\frac{\partial M_x}{\partial x} \, dy \, dx \quad - \quad Q_x \, dy \, dx \quad - \quad \frac{\partial M_{xy}}{\partial y} \, dx \, dy \quad = \quad 0$$

from which

$$\frac{\partial M_x}{\partial x} \quad - \quad \frac{\partial M_{xy}}{\partial y} \quad = \quad Q_x \qquad\qquad (2.7)$$

Similarly, the equilibrium equation of the moments in y-direction yields:

$$\frac{\partial M_y}{\partial y} \quad - \quad \frac{\partial M_{xy}}{\partial x} \quad = \quad Q_y \qquad\qquad (2.8)$$

The equilibrium equation of the vertical forces in z- direction is

$$p \, dx \, dy \quad + \quad \frac{\partial Q_x}{\partial x} \, dy \, dx \quad + \quad \frac{\partial Q_y}{\partial y} \, dx \, dy \quad = \quad 0$$

from which

$$\frac{\partial Q_x}{\partial x} + \frac{\partial Q_y}{\partial y} = -p \tag{2.9}$$

in which
p = the load intensity.

The governing differential equations for the deflection and shearing forces can now be obtained by substituting the moment expressions (2.1) to (2.3) into the equilibrium

equations (2.7) to (2.9). For an orthotropic laminar plate having isotropic facings, these equations are derived by many others [5, 9, 23, 30] as:

$$(1 - A_y \frac{\partial^2}{\partial x^2} - A_x \frac{\partial^2}{\partial y^2}) \nabla\nabla w = \frac{1}{D} (1 - D_2 \frac{\partial^2}{\partial x^2} - D_1 \frac{\partial^2}{\partial y^2}$$

$$+ \frac{2 A_x A_y}{1-\nu} \nabla\nabla) \, p$$

$$(\frac{2 A_x}{1-\nu} - \frac{1+\nu}{1-\nu} A_y) \frac{\partial^2 Q_x}{\partial x^2} + A_x \frac{\partial^2 Q_x}{\partial y^2} - Q_x =$$

$$D \frac{\partial}{\partial x} \nabla w + \frac{1+\nu}{1-\nu} A_y \frac{\partial p}{\partial x}$$

$$(\frac{2 A_y}{1-\nu} - \frac{1+\nu}{1-\nu} A_x) \frac{\partial^2 Q_y}{\partial y^2} + A_y \frac{\partial^2 Q_y}{\partial x^2} - Q_y =$$

$$D \frac{\partial}{\partial y} \nabla w + \frac{1+\nu}{1-\nu} A_x \frac{\partial p}{\partial y}$$

in which

$$\nabla = \frac{\partial^2}{\partial x^2} + \frac{\partial^2}{\partial y^2}$$

$$D' = D'_x = D'_y$$

$$= E_f h^2 \left(\frac{1}{t_1} + \frac{1}{t_2}\right)^{-1}$$

$$D_{xy} = \frac{E_f h^2}{2(1+\nu)} \left(\frac{1}{t_1} + \frac{1}{t_2}\right)^{-1}$$

$$D_\nu = \frac{E_f h^2}{\nu} \left(\frac{1}{t_1} + \frac{1}{t_2}\right)^{-1}$$

$$D = D' \left[1 - \left(\frac{D'}{D_\nu}\right)^2\right]^{-1}$$

$$A_x = \frac{D_{xy}}{S_x}$$

$$A_y = \frac{D_{xy}}{S_y}$$

$$D_1 = A_x + \frac{2A_y}{1-\nu}$$

$$D_2 = A_y + \frac{2A_x}{1-\nu}$$

For the particular case of an isotropic laminar plate with similar facings the governing equations become:

$$\nabla\nabla w \;=\; \frac{1}{D}\, p \;-\; \frac{1}{S}\, \nabla p \tag{2.10}$$

$$\frac{1-\nu}{2}\;\nabla\gamma_x \;-\; \frac{S}{D}\,\gamma_x \;=\; \frac{\partial}{\partial x}\,\nabla w \;+\; \frac{1+\nu}{2S}\,\frac{\partial p}{\partial x} \tag{2.11}$$

$$\frac{1-\nu}{2}\;\nabla\gamma_y \;-\; \frac{S}{D}\,\gamma_y \;=\; \frac{\partial}{\partial y}\,\nabla w \;+\; \frac{1+\nu}{2S}\,\frac{\partial p}{\partial y} \tag{2.12}$$

If a solution of these equations is found that satisfies the boundary conditions of the plate, the moments can be readily obtained from Equs. (2.1) to (2.3). It is of interest to note that, when $G_c = \infty$, Equ. (2.10) reduces to the governing differential equation $\nabla\nabla w = p/D$ in the theory of homogenous classic plates.

## 2.4 SOLUTIONS FOR THE GOVERNING DIFFERENTIAL EQUATIONS

Mathematically, the governing equations of laminar plates are classified as linear partial differential equations with constant coefficients. Equation (2.10) are of fourth order. Analytical solutions for these equations are often of Navier's or Levy's types. Navier's method, in terms of double trigonometric series, is adopted throughout this chapter, even though Levy's solution, in the form of a single series, converges faster. Many reasons justify this decision. First, it provides a means to solve Equs. (2.11) and (2.12); second, sufficiently accurate results can still be obtained by using more terms, and finally considerable mathematical simplicity is achieved in the solution of simply supported plates.

Consider now the simply supported plate in Fig. 2.1. Three boundary conditions at each edge of the plate were admitted by the sixth order governing differential equations for the deflection and are commonly found in practice [3, 5, 9, 21, 23, 30]. These conditions are:

$$w = o, \; M_x = o, \; \gamma_y = o \qquad \text{for} \qquad x = o \text{ and } x = a$$

and $\hfill$ (2.13)

$$w = 0, M_y = 0, \gamma_x = 0 \qquad \text{for} \qquad y = 0 \text{ and } y = b$$

in which
　　$a, b$　 = the plate dimensions parallel to x- and y-directions, respectively.

The deflected surface of the plate satisfies the condition $w = 0$ in Equ. (2.13) is expressed in double trigonometric series as:

$$w = \sum_{m=1}^{\infty} \sum_{n=1}^{\infty} W_{mn} \sin \alpha_m x \sin \beta_n y \qquad (2.14)$$

in which

$$\alpha_m = \frac{m\pi}{a}$$

$$\beta_n = \frac{n\pi}{b}$$

$$W_{mn} = \text{the coefficient of the double Fourier expansion of the deflection.}$$

The applied load can be similarly expressed in double series as

$$p(x,y) = \sum_{m=1}^{\infty} \sum_{n=1}^{\infty} P_{mn} \sin \alpha_m x \sin \beta_n y \qquad (2.15)$$

in which

    $P_{mn}$   = same meaning as $W_{mn}$ but for the applied load.

Substituting Equs. (2.14) and (2.15) in the governing equation (2.10), an algebraic equation is obtained, from which the unknown $W_{mn}$ is determined as

$$W_{mn} = P_{mn} \; \frac{(1/D) + [(\alpha_m^2 + \beta_n^2)/S]}{(\alpha_m^2 + \beta_n^2)^2} \tag{2.16}$$

The expression for the plate deflection is now obtained by summing the individual terms in Equ. (2.14) and by making use of Equ. (2.16).

$$w = \sum_{m=1}^{\infty} \sum_{n=1}^{\infty} P_{mn} \; \frac{(1/D) + [(\alpha_m^2 + \beta_n^2)/S]}{(\alpha_m^2 + \beta_n^2)^2} \; \sin \alpha_m x \, \sin \beta_n y \tag{2.17}$$

This expression reveals that the deflection of a laminar plate is composed of two components: the bending deflection which occurs in a plate rigid in shear, and the shear deflection which occurs in a plate infinitely stiff in bending. These two components were separately obtained by Plantema [21] based on the partial deflection theory.

Returning to the plate in Fig. 2.1, expressions for the shear deformations can be obtained now from Equ. (2.14) and the results are in agreement with those given in [8, 20, 22]. Thus,

$$\gamma_x = \sum_{m=1}^{\infty} \sum_{n=1}^{\infty} \gamma_{xmn} \cos \alpha_m x \, \sin \beta_n y \tag{2.18}$$

$$\gamma_y = \sum_{m=1}^{\infty} \sum_{n=1}^{\infty} \gamma_{ymn} \sin \alpha_m x \, \cos \beta_n y \tag{2.19}$$

in which

    $\gamma_{xmn}, \gamma_{ymn}$   =   the coefficients of the double Fourier expansion of the shearing deformations.

It should be noted that both Equs. (2.18) and (2.19) satisfy the conditions $\gamma = o$ in Equ. (2.13); and together with Equ. (2.14), they satisfy the conditions $M = o$ along the edges.

Making use of Equs. (2.14), (2.18), and (2.19), the coefficients $\gamma_{xmn}$ and $\gamma_{ymn}$ can be readily determined from Equs. (2.11) and (2.12) respectively. The results are

$$\gamma_{xmn} = \frac{m\pi\, P_{mn}}{a S(\alpha_m^2 + \beta_n^2)} \tag{2.20}$$

and

$$\gamma_{ymn} = \frac{n\pi\, P_{mn}}{b S(\alpha_m^2 + \beta_n^2)} \tag{2.21}$$

The final expressions for the shear forces in a simply supported laminar plate subjected to transverse load are obtained as:

$$Q_x = \sum_{m=1}^{\infty} \sum_{n=1}^{\infty} \frac{m\pi\, P_{mn}}{a(\alpha_m^2 + \beta_n^2)} \cos \alpha_m x \sin \beta_n y \tag{2.22}$$

$$Q_y = \sum_{m=1}^{\infty} \sum_{n=1}^{\infty} \frac{n\pi\, P_{mn}}{b(\alpha_m^2 + \beta_n^2)} \sin \alpha_m x \cos \beta_n y \tag{2.23}$$

## 2.5  MOMENTS IN SIMPLY SUPPORTED COMPOSITE PLATES

Having solved the governing differential equations (2.10) to (2.12), the bending and twisting moments at any point of a simply supported plate can be obtained now from Equs. (2.1) to (2.3). This is by direct substitution of Equs. (2.17), (2.22), and (2.23) into the moment expressions.

$$M_x = \sum_{m=1}^{\infty} \sum_{n=1}^{\infty} P_{mn} \frac{(\alpha_m^2 + \nu\beta_n^2)}{(\alpha_m^2 + \beta_n^2)^2} \sin \alpha_m x \ \sin \beta_n y \tag{2.24}$$

$$M_y = \sum_{m=1}^{\infty} \sum_{n=1}^{\infty} P_{mn} \frac{(\beta_n^2 + \nu\alpha_m^2)}{(\alpha_m^2 + \beta_n^2)^2} \sin \alpha_m x \ \sin \beta_n y \tag{2.25}$$

$$M_{xy} = (1-\nu) \sum_{m=1}^{\infty} \sum_{n=1}^{\infty} \frac{\pi^2 P_{mn}}{ab} \cdot \frac{mn}{(\alpha_m^2 + \beta_n^2)^2} \cos \alpha_m x \cos \beta_n y \quad (2.26)$$

## 2.6 PRACTICAL FORMULAS FOR FORCES AND STRESSES

Existing expressions [3, 21, 31] for evaluating the deflection and stresses in simply supported plates are more oriented towards mathematics than practical engineering. For example, Allen's formulas [3] involve many complex calculations, whereas Yen's results [31] are restricted to square plates. In addition, only uniform loading, in the former study, and concentrated loading, in the latter one, have been considered.

In the following, formulas are developed for simply supported plates under transverse load of five types: uniformly distributed, hydrostatic pressure, partial load on rectangular area, concentrated load, and strip load uniformaly distributed across the plate width. The formulas developed are simpler than previous ones [3, 21, 31]. For this purpose, consider the simply supported laminar plate in Fig. 2.1. The deflection and bending moments at the plate center are obtained from Equs. (2.17), (2.24), and (2.25) as follows

$$(w)_{\substack{x=a/2 \\ y=b/2}} = - \sum_{m=1,3,\ldots}^{\infty} \sum_{n=1,3,\ldots}^{\infty} (-1)^{(m+n)/2} \frac{P_{mn}}{C} [\frac{1}{D} + \frac{1}{S} (\alpha_m^2 + \beta_n^2 )] \quad (2.27)$$

$$(M_x)_{\substack{x=a/2 \\ y=b/2}} = - \sum_{m=1,3,\ldots}^{\infty} \sum_{n=1,3,\ldots}^{\infty} (-1)^{(m+n)/2} \frac{P_{mn}}{C} (\alpha_m^2 + \nu\beta_n^2 ) \quad (2.28)$$

$$(M_y)_{\substack{x=a/2 \\ y=b/2}} = - \sum_{m=1,3,\ldots}^{\infty} \sum_{n=1,3,\ldots}^{\infty} (-1)^{(m+n)/2} \frac{P_{mn}}{C} (\beta_n^2 + \nu\alpha_m^2 ) \quad (2.29)$$

The twisting moment is maximum at the plate corners and is obtained from Equ. (2.26) as:

$$(M_{xy})_{\substack{x=0 \\ y=0}} = (1-\nu) \sum_{m=1}^{\infty} \sum_{n=1}^{\infty} \frac{\pi^2}{ab} \frac{P_{mn}}{C} mn \qquad (2.30)$$

The concentrated shear force at the plate corners can be obtained from Equ. (2.30) [11, 28, 29]. The shear forces are calculated at the mid edges from Equs. (2.22) and (2.23) as:

$$(Q_x)_{\substack{x=0 \\ y=b/2}} = - \sum_{m=1}^{\infty} \sum_{n=1,3,\ldots}^{\infty} (-1)^{(n+1)/2} \frac{P_{mn}}{\sqrt{C}} \frac{m\pi}{a} \qquad (2.31)$$

$$(Q_y)_{\substack{x=a/2 \\ y=0}} = - \sum_{m=1,3,\ldots}^{\infty} \sum_{n=1}^{\infty} (-1)^{(m+1)/2} \frac{P_{mn}}{\sqrt{C}} \frac{n\pi}{b} \qquad (2.32)$$

in which

$$C = (\alpha_m^2 + \beta_n^2)^2$$

where the coefficient $P_{mn}$ of Fourier expansion of the applied load, may be obtained from the following equation [28, 29]:

$$P_{mn} = \frac{4}{ab} \int_o^a \int_o^b p(x,y) \sin \alpha_m x \sin \alpha_n y \, dxdy \qquad (2.33)$$

in which
   $p(x,y)$ = function representing the external applied load.

Equations (2.27) to (2.32) furnish the analytical formulas from which the results corresponding to individual loading are derived. Detailed derivation is presented for the case of uniform loading, and only the final results are given for the other four cases.

### 2.6.1   UNIFORM DISTRIBUTED LOAD

The applied load in this case is uniformly distributed of intensity $P_o$ (Fig. 2.3(a)):

$$p(x,y) = P_o$$

Substituting of Equ. (2.34) into Equ. (2.33) and performing the double integration yields:

$$P_{mn} = \begin{cases} \dfrac{16\,P_o}{\pi^2 mn} & \text{for odd m and n} \\ \\ 0 & \text{otherwise} \end{cases} \qquad (2.35)$$

The central deflection is obtained now by substituting Equ. (2.35) into Equ. (2.27)

$$(w)_{\substack{x=a/2 \\ y=b/2}} = - \sum_{m=1,3,\ldots}^{\infty} \sum_{n=1,3,\ldots}^{\infty} (-1)^{(m+n)/2}\, \frac{16\,P_o}{\pi^2 Cmn}\, (\frac{1}{D} +$$

$$\frac{\alpha_m^2 + \beta_n^2}{S}) \qquad (2.36)$$

In a more compact form, the central deflection equation can be written as:

$$(w)_{\substack{x=a/2 \\ y=b/2}} = \frac{P_o a^4}{D}\, K_{wb} + \frac{P_o a^2}{S}\, K_{ws} \qquad (2.37)$$

in which

$$K_{wb} = \sum_{m=1,3,\ldots}^{\infty} \sum_{n=1,3,\ldots}^{\infty} \frac{-16 \, (-1)^{(m+n)/2}}{\pi^6 mnC'^2}$$

$$\left.\begin{matrix} \\ \\ \\ \end{matrix}\right\} \quad (2.38)$$

$$K_{ws} = \sum_{m=1,3,\ldots}^{\infty} \sum_{n=1,3,\ldots}^{\infty} \frac{-16 \, (-1)^{(m+n)/2}}{\pi^4 mnC'}$$

$$C' = m^2 + R^2 n^2$$

It can be seen from this equation that two terms are to be added: the bending and shear components of the deflection, and each involves two factors. The first contains parameters reflecting the material properties and plate geometry and the second the aspect ratio of the plate. To facilitate the use of Equ. (2.37), numerical results are calculated for $K_{wb}$ and $K_{ws}$ and presented in the second and third columns of Table A.1. A range of one to five for the plate aspect ratio was considered.

Concerning the bending and twisting moments produced in a simply supported plate due to a uniformly distributed load, a similar procedure as before is used. This is by substituting Equ. (2.35) in Equs. (2.28), (2.29), and (2.30) from which it follows that:

$$(M_x)_{\substack{x=a/2 \\ y=b/2}} = p_0 a^2 \, K_{mx}$$

$$(M_y)_{\substack{x=a/2 \\ y=b/2}} = p_0 a^2 \, K_{my} \qquad (2.40)$$

$$(M_{xy})_{\substack{x=0 \\ y=0}} = - p_0 a^2 R \, K_{mxy} \qquad (2.41)$$

in which

$$K_{mx} = \sum_{m=1,3,\ldots}^{\infty} \sum_{n=1,3,\ldots}^{\infty} (-1)^{(m+n)/2} \frac{-16}{\pi^4 mnC'^2} \left( m^2 + \nu R^2 n^2 \right)$$

$$\left.\begin{matrix} \\ \\ \\ \\ \\ \end{matrix}\right\} \quad (2.42)$$

$$K_{my} = \sum_{m=1,3,\ldots}^{\infty} \sum_{n=1,3,\ldots}^{\infty} (-1)^{(m+n)/2} \frac{-16}{\pi^4 mnC'^2} \left( R^2 n^2 + \nu m^2 \right)$$

$$K_{mxy} = \sum_{m=1,3,\ldots}^{\infty} \sum_{n=1,3,\ldots}^{\infty} \frac{16(\nu-1)}{\pi^4 c_i^2}$$

Detailed investigations of the convergence of the series in Equ. (2.42) are reported in Ref. [3, 11, 28, 29]. Values for the factors $K_{mx}$, $K_{my}$, and $K_{mxy}$ are obtained by taking the sums of the series in Equ. (2.42) up to and including the term m=n=100. The results are tabulated in Table A.1.

Stresses are sometimes of interest in many problems. The inplane stresses in the outer laminae may be obtained from the results (2.39), (2.40), and (2.41). For example, the normal stress $\sigma_x$ and $\sigma_y$ are calculated by:

$$\sigma_x = \frac{M_x}{ht_f}$$

$$= \frac{p_o a^2}{ht_f} K_{mx} \tag{2.43}$$

and

$$\sigma_y = \frac{M_y}{ht_f}$$

$$= \frac{p_o a^2}{ht_f} K_{my} \tag{2.44}$$

The shear stress $\tau_{xy}$ is obtained from Equ. (2.41) as:

$$\tau_{xy} = \frac{M_{xy}}{ht_f}$$

$$= - \frac{p_o a^2 R}{h t_f} K_{mxy} \tag{2.45}$$

Equs. (2.43), (2.44), and (2.45) represent the peak inplane stresses in the faces of a simply supported plate subjected to a uniformly distributed load.

It is of interest to observe that the expressions for the stresses do not include $G_c$, nor any other term which refers to the shear stiffness of the plate. Indeed, not only are the stresses in the faces independent of the shear stiffness, but also the results in Equs. (2.43), (2.44), and (2.45) are identical to those determined by the theory of homogenous plates [28, 29]. This is true due to the structural manner in which a simply supported plate deforms. Under the applied transverse load, the thin laminae of a simply supported plate undergo uniform extension or contractions as they bend about the middle plane of the whole plate. In this way, the inplane stresses are uniformly distributed across the thicknesses. In addition, the plate undergoes shear strains which correspond to an additional transverse deflection. The faces accomodate this extra deflection by bending about their own middle planes, as well as by displacing vertically only. This implies that, while shear deformations are taking place, the thin laminae of a simply supported plate do not undergo stretching or contraction, thus they retain their bending stresses unaltered.

Returning to the plate in Fig. 2.3(a), the shear forces in the simply supported plate under uniformly distributed load are obtained from Equs. (2.31) and (2.32) as:

$$(Q_x)_{\substack{x=0 \\ y=b/2}} = p_o a\, K_{Qx} \tag{2.46}$$

$$(Q_y)_{\substack{x=a/2 \\ y=0}} = p_o a\, R\, K_{Qy} \tag{2.47}$$

in which

$$K_{Qx} = \sum_{m=1,3,\ldots}^{\infty} \sum_{n=1,3,\ldots}^{\infty} (-1)^{(n+1)/2} \frac{-16}{\pi^3 n\, C'}$$

$$K_{Qy} = \sum_{m=1,3,\ldots}^{\infty} \sum_{n=1,3,\ldots}^{\infty} (-1)^{(m+1)/2} \frac{-16}{\pi^3 m\, C'}$$

$$(2.48)$$

Numerical values for the coefficients $K_{Qx}$ and $K_{Qy}$ are evaluated and shown in Table A.1. The shear stresses can be obtained from:

$$\tau_{xz} = \frac{Q_x}{h}$$

$$= \frac{P_o\, a}{h} K_{Qx} \qquad\qquad (2.49)$$

$$\tau_{yz} = \frac{P_o\, a\, R}{h} K_{Qy} \qquad\qquad (2.50)$$

While the transverse shear stress $\tau_{xz}$ is greatest at the middle of the sides of length b (e.g. at $x = 0$, a and $y = b/2$), the transverse shear stress $\tau_{yz}$ is greatest at the middle of the sides of length a (e.g. $x = a/2$, $y = 0$ and b).

In summary, a uniformly distributed load can be expressed mathematically by Equ. (2.34) which together with Equ. (2.33) leads to an expression for the coefficient of Fourier expansion of the load. Having determined this coefficient, the deflection, shears, moments, and stresses produced due to the applied load are calculated readily from Equs. (2.27) to (2.32). The formulas developed are simplified and numerical values are also tabulated for a wide range of parameters.

## 2.6.2 HYDROSTATIC PRESSURE

The hydrostatic pressure shown in Fig. 2.3(b) can be represented by the equation

$$p(x,y) = P_o x/a$$

in which
$p_o$ = the load intensity at $x = a$.

In this case, the coefficients of Fourier expansion for the load are obtained as

$$P_{mn} = 8 \, (-1)^{m+1} p_o / \pi^2 mn$$

where   $m = 1, 2, 3, \ldots$                    and $n = 1, 3, 5, \ldots$

Similar to the case of uniform distributed load, the expressions for the deflection, bending and twisting moments, and shears produced in a simply supported plate due to the hydrostatic pressure, can also be represented by Equs. (2.37), (2.39), (2.40), (2.41), (2.46), and (2.47). However, the factors in these equations are given by

$$
\left.
\begin{aligned}
K_{wb} &= \sum_{m=1,3,\ldots}^{\infty} \sum_{n=1,3,\ldots}^{\infty} (-1)^{(3m+n)/2} \frac{8}{\pi^6 mnC'^2} \\[2em]
K_{ws} &= \sum_{m=1,3,\ldots}^{\infty} \sum_{n=1,3,\ldots}^{\infty} (-1)^{(3m+n)/2} \frac{8}{\pi^4 mnC'} \\[2em]
K_{mx} &= \sum_{m=1,3,\ldots}^{\infty} \sum_{n=1,3,\ldots}^{\infty} (-1)^{(3m+n)/2} \frac{8}{\pi^4 mnC'^2} ( \\
&\qquad\qquad m^2 + \nu R^2 n^2 )
\end{aligned}
\right\} \quad (2.53)
$$

$$K_{my} = \sum_{m=1,3,\ldots}^{\infty} \sum_{n=1,3,\ldots}^{\infty} (-1)^{(3m+n)/2} \frac{8}{\pi^4 mnC'^2} ($$

$$R^2 n^2 + \nu m^2 )$$

$$K_{mxy} = \sum_{m=1}^{\infty} \sum_{n=1,3,\ldots}^{\infty} (-1)^m \frac{8\ (1-\nu)}{\pi^4 \ C'^2}$$

$$K_{Qx} = \sum_{m=1}^{\infty} \sum_{n=1,3,\ldots}^{\infty} (-1)^{(2m+n+1)/2} \frac{8}{\pi^3\ n\ C'}$$

$$K_{Qy} = \sum_{m=1,3,\ldots}^{\infty} \sum_{n=1,3,\ldots}^{\infty} (-1)^{(3m+1)/2} \frac{8}{\pi^3\ m\ C'}$$

Numerical values for these factors are tabulated in Table A.2.

### 2.6.3  PARTIAL LOAD ON A RECTANGULAR AREA

In the case of partial load on a rectangular area, as shown in Fig. 2.3(c), the loading function is

$$p\ (x,y) = \begin{cases} p_0 & \text{for } x = (\xi - \frac{c}{2}) \text{ to } x = (\xi + \frac{c}{2}) \\ & \text{and } y = (\eta - \frac{d}{2}) \text{ to } y = (\eta + \frac{d}{2}) \\ \\ 0 & \text{otherwise} \end{cases}$$

$$(2.54)$$

in which

$p_0$ = the load intensity;

c x d = the loaded area;

$\xi, \eta$ = x- and y-coordinates of the loaded area's center, respectively.

The coefficient of Fourier expansion of the partial load is obtained from Equ. (2.33) as:

$$P_{mn} = \frac{16}{\pi^2} \frac{P_o}{mn} \sin \frac{m\pi\xi}{a} \sin \frac{n\pi\eta}{b} \sin \frac{m\pi c}{2a} \sin \frac{n\pi d}{2b} \qquad (2.55)$$

for all integers of m and n.

The factors in Equs. (2,37), (2.39), (2.40), (2.41), (2.46), and (2.47) for the present loading case are given by

$$K_{wb} = \sum_{m=1,3,\dots}^{\infty} \sum_{n=1,3,\dots}^{\infty} (-1)^{(m+n)/2} \frac{-16\ C''}{\pi^6 mn C'^2}$$

$$K_{ws} = \sum_{m=1,3,\dots}^{\infty} \sum_{n=1,3,\dots}^{\infty} (-1)^{(m+n)/2} \frac{-16\ C''}{\pi^4 mn C'}$$

$$K_{mx} = \sum_{m=1,3,\dots}^{\infty} \sum_{n=1,3,\dots}^{\infty} (-1)^{(m+n)/2} \frac{-16\ C''}{\pi^4 mn C'^2} ( m^2 + \nu R^2 n^2 )$$

$$K_{my} = \sum_{m=1,3,\dots}^{\infty} \sum_{n=1,3,\dots}^{\infty} (-1)^{(m+n)/2} \frac{-16\ C''}{\pi^4 mn C'^2} ( R^2 n^2 + \nu m^2 )$$

$$K_{mxy} = \sum_{m=1}^{\infty} \sum_{n=1}^{\infty} \frac{16(\nu-1)C''}{\pi^4 C'^2}$$

$$K_{Qx} = \sum_{m=1}^{\infty} \sum_{n=1,3,\dots}^{\infty} (-1)^{(n+1)/2} \frac{-16\ C''}{\pi^3 n\ C'}$$

$$K_{Qy} = \sum_{m=1,3,\dots}^{\infty} \sum_{n=1}^{\infty} (-1)^{(m+1)/2} \frac{-16\ C''}{\pi^3 m\ C'}$$

$$(2.56)$$

in which

$$C'' = \sin \frac{m\pi\xi}{a} \, \sin \frac{n\pi\eta}{b} \, \sin \frac{m\pi c}{2a} \, \sin \frac{n\pi d}{2b}$$

In addition to the aspect ratio, the position of the load on the plate is another parameter in Equ. (2.56); but due to symmetry only one quarter of the plate surface is considered. This quarter is divided into 3×3 rectangular mesh, each of size (a/6) × (b/6); and the loaded area's center is fixed at the mesh points. The ratio of loaded area to plate surface is chosen as follows

c/a = d/b = 1/4

hence,

(c × d)/(a × b) = 1/16

Based on this choice, numerical values for the factors in Equ. (2.56) are evaluated for each position of the partial load and the results are presented in Tables A.3 to A.11.

## 2.6.4  CONCENTRATED LOAD

Concentrated load is a particular case of the partial load in which the loaded area is diminishingly small. Thus, the coefficient of Fourier's expansion for the concentrated load is obtained from Equ. (2.55) by substituting $p_o = P/cd$, where P is the concentrated load shown in Fig. 2.3(d). Letting the loaded area approach zero by permitting c → o and d → o, yields:

$$P_{mn} = \frac{4P}{ab} \, \sin \frac{m\pi\xi}{a} \, \sin \frac{n\pi\eta}{b} \qquad\qquad (2.57)$$

for all integers of m and n.

The procedure to follow in developing the practical formulas for a simply supported plate subjected to a concentrated load is not different from that given in the previous section; however, the formulas obtained are given by:

$$(w)_{\substack{x=a/2 \\ y=b/2}} = \frac{P\,a^2\,R}{D}\,K_{wb} + \frac{P\,R}{S}\,K_{ws} \tag{2.58}$$

$$(M_x)_{\substack{x=a/2 \\ y=b/2}} = P\,R\,K_{mx} \tag{2.59}$$

$$(M_y)_{\substack{x=a/2 \\ y=b/2}} = P\,R\,K_{my} \tag{2.60}$$

$$(M_{xy})_{\substack{x=0 \\ y=0}} = -P\,R^2\,K_{mxy} \tag{2.61}$$

$$(Q_x)_{\substack{x=a/2 \\ y=0}} = \frac{P\,R}{a}\,K_{Qx} \tag{2.62}$$

$$(Q_y)_{\substack{x=0 \\ y=b/2}} = \frac{P\,R^2}{a}\,K_{Qy} \tag{2.63}$$

in which

$$K_{wb} = \sum_{m=1,3,\dots}^{\infty}\ \sum_{n=1,3,\dots}^{\infty} (-1)^{(m+n)/2}\,\frac{-4}{\pi^4}\,\frac{C'''}{C'^2}$$

$$K_{ws} = \sum_{m=1,3,\dots}^{\infty}\ \sum_{n=1,3,\dots}^{\infty} (-1)^{(m+n)/2}\,\frac{-4}{\pi^2}\,\frac{C'''}{C'}$$

$$K_{mx} = \sum_{m=1,3,\dots}^{\infty}\ \sum_{n=1,3,\dots}^{\infty} (-1)^{(m+n)/2}\,\frac{-4}{\pi^2}\,\frac{C'''}{C'^2}\,(\ m^2 + \nu R^2 n^2\ )$$

$$K_{my} = \sum_{m=1,3,\ldots}^{\infty} \sum_{n=1,3,\ldots}^{\infty} (-1)^{(m+n)/2} \frac{-4}{\pi^2} \frac{C'''}{C'^2} ($$

$$R^2 n^2 + \nu m^2)$$                    (2.64)

$$K_{mxy} = \sum_{m=1}^{\infty} \sum_{n=1}^{\infty} \frac{4(\nu-1)mnC'''}{\pi^2 C'^2}$$

$$K_{Qx} = \sum_{m=1}^{\infty} \sum_{n=1,3,\ldots}^{\infty} (-1)^{(n+1)/2} \frac{-4m}{\pi} \frac{C'''}{C'}$$

$$K_{Qy} = \sum_{m=1,3,\ldots}^{\infty} \sum_{n=1}^{\infty} (-1)^{(m+1)/2} \frac{-4n}{\pi} \frac{C'''}{C'}$$

where

$C'''$    = $\sin (m\pi\xi/a) \sin (n\pi\eta/b)$

$\xi, \eta$    = x- and y-coordinates of the concentrated load's position on the plate respectively, as shown in Fig. 2.3(d).

Numerical values for the factors in Equ. (2.64) are given in Tables A.12 to A.20.

### 2.6.5  LINE LOAD

Line load is another particular case of the partial load which was considered in section 2.6.3. As shown in Fig. 2.3(e), the load is uniformly distributed across the plate width with a constant intensity. In this case, the coefficient of Fourier expansion of the load is obtained by substituting

$\eta = b/2$

and

$$c = b$$

into Equ. (2.55) and then by permitting c→o, it follows that:

$$P_{mn} = (8 \, p_0/\pi an) \sin (m\pi\xi/a)$$

in which
$p_0$ = the load intensity per unit length of the side b;
$\xi$ = the x-coordinate of the line load (Fig. 2.3(e)).

The formulas for the deflection, moments, and shears of a simply supported sandwich plate subjected to the line load are obtained as:

$$(w)_{\substack{x=a/2 \\ y=b/2}} = \frac{2 \, p_0 a^3}{D} K_{wb} + \frac{2 \, p_0 a}{S} K_{ws} \qquad (2.66)$$

$$(M_x)_{\substack{x=a/2 \\ y=b/2}} = 2 \, p_0 a \, K_{mx} \qquad (2.67)$$

$$(M_y)_{\substack{x=a/2 \\ y=b/2}} = 2 \, p_0 a \, K_{my} \qquad (2.68)$$

$$(M_{xy})_{\substack{x=0 \\ y=0}} = - \, 2 \, p_0 R a \, K_{mxy} \qquad (2.69)$$

$$(Q_x)_{\substack{x=a/2 \\ y=0}} = 2 \, p_0 \, K_{Qx} \qquad (2.70)$$

$$(Q_y)_{\substack{x=0 \\ y=b/2}} = 2 \, p_0 R \, K_{Qy} \qquad (2.71)$$

in which

$$K_{wb} = \sum_{m=1,3,\ldots}^{\infty} \sum_{n=1,3,\ldots}^{\infty} (-1)^{(m+n)/2} \frac{-4}{\pi^5 n C'^2} \sin \frac{m\pi\xi}{a}$$

$$K_{ws} = \sum_{m=1,3,\ldots}^{\infty} \sum_{n=1,3,\ldots}^{\infty} (-1)^{(m+n)/2} \frac{-4}{\pi^3 n C'} \sin \frac{m\pi\xi}{a}$$

$$K_{mx} = \sum_{m=1,3,\ldots}^{\infty} \sum_{n=1,3,\ldots}^{\infty} (-1)^{(m+n)/2} \frac{-4}{\pi^3 n C'^2} ( m^2 + \nu R^2 n^2) \sin \frac{m\pi\xi}{a}$$

$$K_{my} = \sum_{m=1,3,\ldots}^{\infty} \sum_{n=1,3,\ldots}^{\infty} (-1)^{(m+n)/2} \frac{-4}{\pi^3 n C'^2} ( R^2 n^2 + \nu m^2) \sin \frac{m\pi\xi}{a}$$

$$K_{mxy} = \sum_{m=1}^{\infty} \sum_{n=1}^{\infty} \frac{4(\nu-1)m}{\pi^3 C'^2} \sin \frac{m\pi\xi}{a}$$

$$K_{Qx} = \sum_{m=1}^{\infty} \sum_{n=1,3,\ldots}^{\infty} (-1)^{(n+1)/2} \frac{-4m}{\pi^2 n C'} \sin \frac{m\pi\xi}{a}$$

$$K_{Qy} = \sum_{m=1,3,\ldots}^{\infty} \sum_{n=1}^{\infty} (-1)^{(m+1)/2} \frac{-4}{\pi^2 C'} \sin \frac{m\pi\xi}{a}$$

$$\left.\right\} (2.72)$$

These factors are to be evaluated for different loading positions, however, due to the symmetry about the central axis x = a/2, only one half of the plate surface is considered. This half is divided into five strips each of a width equal to one-tenth of the side length a. The line load is applied on the lines where each two adjacent strips are intersected, and the corresponding numerical values for the factors in Equ. (2.72) are given in Tables A.21 to A.25.

In summary, simply supported laminar plates subjected to various types of lateral loads are solved by applying Navier's method. To facilitate the use of the solutions developed and to reduce the amount of work involved, practical formulas are obtained from which the deflection, moments, and shears can be calculated. Two factors are involved in the formulas. The first contains parameters reflecting the material properties and the panel geometry and the second the aspect ratio and the load position. Numerical values for the latter factor can be obtained from Tables A.1 to A.25.

## 2.7 SIMPLY SUPPORTED COMPOSITE PLATES BY THE FINITE DIFFERENCE METHOD

Analytic solutions to the governing differential equations of many plate problems cannot be easily found [28]. However, for most practical purposes, acceptable results can be obtained by numerical treatment of the governing differential equations. Among the numerical techniques presently available, the finite difference method is one of the most general. In applying this method, the derivatives in the differential equation under consideration are replaced, in general, by difference quantities at some located points that form a reference network called finite difference mesh, and consequently, the governing differential equations are transformed into a set of algebraic simultaneous equations. Solving these equations yields approximate values for the parameter being described by the governing equation under consideration, for instance a plate deflection.

The applications of two finite difference methods are presented here [28]. These are: the higher approximation method, and the funicular polygon method. The stencil of the former method is shown in Fig. 2.4 whereas for the latter method, it is shown in Fig. 2.5.

The finite difference method is used now in conjunction with the partial deflection theory to determine the deflection of a simply supported square plate subjected to three loading types: a uniformly distributed load, a hydrostatic pressure, and a central partial load. The governing differential equations for the bending and shear deflections are obtained from Equ. (2.10) as:

$$\nabla\nabla w_b = p/D \qquad\qquad (2.73)$$

$$\nabla w_s = -p/S \qquad\qquad (2.74)$$

in which

$\nabla$ $\quad = (\partial^2/\partial x^2) + (\partial^2/\partial y^2);$

$w_b, w_s$ = the bending and shear deflections, respectively;

p $\quad\quad$ = the intensity of the applied load on the plate.

The bending deflection $w_b$ is obtained from Equ. (2.73) by applying the funicular polygon method stencil in Fig. 2.5 to the finite difference mesh points. The resulting equations are written in matrix form as:

$$[A] \; \{w_b\} = \{P_b\} \tag{2.75}$$

in which

[A] $\quad\quad$ = a matrix containing the coefficients of the bending deflection ordinates;

$\{P_b\}$ $\quad\quad$ = the joint load vector;

$\{w_b\}$ $\quad\quad$ = a vector containing the bending deflection ordinates at the mesh points.

From Equ. (2.75) it follows that:

$$\{w_b\} = [A]^{-1} \; \{P_b\} \tag{2.76}$$

In a similar manner, the shear deflection $w_s$ is obtained from Equ. (2.74) by applying the higher order stencil in Fig. 2.4 to the finite difference mesh points, and in matrix form the result is:

$$[B] \{w_s\} = \{P_s\} \tag{2.77}$$

in which

[B], $\{P_s\}$, $\{w_s\}$ = same meaning as [A], $\{P_b\}$, and $\{w_b\}$ respectively, but for the shear deflection

from which

$$\{w_s\} = [B]^{-1} \; \{P_s\} \tag{2.78}$$

Consider now, the simply supported square plates shown in Figs. 2.6, 2.7, and 2.8 which show the finite difference meshes and the boundary conditions. The coefficient matrices, the load vectors, and the deflections obtained are given in the next section for each of the loading types.

### 2.7.1 UNIFORMLY LOADED SQUARE PLATE, Fig. 2.6

The coefficient matrices [A], [B], and the load vectors $\{P_b\}$, $\{P_s\}$, in this case are:

$$
[A] = \begin{array}{c} \\ \\ \\ \\ \end{array}
\begin{array}{cccc}
w_{b1} & w_{b2} & w_{b3} & w_{b4} \\
\left[\begin{array}{rrrr}
1872. & -576. & -576. & -32. \\
-1152. & 1872. & -64. & -576. \\
-1152. & -64. & 1872. & -576. \\
-128. & -1152. & -1152. & 1872.
\end{array}\right] & \begin{array}{c} 1 \\ 2 \\ 3 \\ 4 \end{array}
\end{array}
$$

$$
[B] = \begin{array}{cccc}
w_{s1} & w_{s2} & w_{s3} & w_{s4} \\
\left[\begin{array}{rrrr}
-60. & 16. & 16. & 0. \\
32. & -60. & 0. & 16. \\
32. & 0. & -60. & 16. \\
0. & 32. & 32. & -60.
\end{array}\right] & \begin{array}{c} 1 \\ 2 \\ 3 \\ 4 \end{array}
\end{array}
$$

$$
\{P_b\} = \lambda^4 p_o / 144 \; D \begin{Bmatrix} 14880. \\ 17320. \\ 17320. \\ 20160. \end{Bmatrix}
$$

$$
\{P_s\} = -12\lambda^2 p_o / S \begin{Bmatrix} 1. \\ 1. \\ 1. \\ 1. \end{Bmatrix}
$$

in which

$\lambda$ $\doteq$ the mesh width;

$\quad$ = $a/4$ (where $a$ is the side length of the plate);

$p_o$ = the uniform load intensity.

From Equs. (2.76) and (2.78) the deflection coordinates are obtained as:

$$
\begin{Bmatrix} w_{b1} \\ w_{b2} \\ w_{b3} \\ \vdots \end{Bmatrix} = p_o a^4 / 36864. \; D \begin{Bmatrix} 70.43 \\ 97.75 \\ 97.75 \\ \vdots \end{Bmatrix}
$$

$$\left\lfloor w_{b4} \right\rfloor \qquad\qquad \left\lfloor 135.90 \right\rfloor$$

$$\left\{ \begin{array}{c} w_{s1} \\ w_{s2} \\ w_{s3} \\ w_{s4} \end{array} \right\} \;=\; 3p_o a^4/4\ S \quad \left\{ \begin{array}{c} 0.054 \\ 0.070 \\ 0.070 \\ 0.091 \end{array} \right\}$$

## 2.7.2  HYDROSTATIC PRESSURE, Fig. 2.7

Similar to the case of uniform pressure, the finite difference mesh for the present loading type is symmetrical in one direction as shown in Fig. 2.7.  The coefficient matrices and load vectors are:

|   | $w_{b1}$ | $w_{b2}$ | $w_{b3}$ | $w_{b4}$ | $w_{b5}$ | $w_{b6}$ |   |
|---|---|---|---|---|---|---|---|
| $[A] =$ | 1800. | -576. | 72. | -608. | -32. | 32. | 1 |
|  | -576. | 1872. | -576. | -32. | -576. | -32. | 2 |
|  | 72. | -576. | 1800. | 32. | -32. | -608. | 3 |
|  | -1216. | -64. | 64. | 1800. | -576. | 72. | 4 |
|  | -64. | -1152. | -64. | -576. | 1872. | -576. | 5 |
|  | 64. | -64. | -1216. | 72. | -576. | 1800. | 6 |

|   | $w_{s1}$ | $w_{s2}$ | $w_{s3}$ | $w_{s4}$ | $w_{s5}$ | $w_{s6}$ |   |
|---|---|---|---|---|---|---|---|
| $[B] =$ | -59. | 16. | -1. | 16. | 0. | 0. | 1 |
|  | 16. | -60. | 16. | 0. | 16. | 0. | 2 |
|  | -1. | 16. | -59. | 0. | 0. | 16. | 3 |
|  | 32. | 0. | 0. | -59. | 16. | -1. | 4 |
|  | 0. | 32. | 0. | 16. | -60. | 16. | 5 |
|  | 0. | 0. | 32. | -1. | 16. | -59. | 6 |

$$\{P_b\} \;=\; p_o\lambda^4/576\ D \quad \left\{ \begin{array}{c} 17570. \\ 34650. \\ 41970. \\ 20450. \\ 42730. \end{array} \right\}$$

$$\left\{\begin{array}{c} 48850. \end{array}\right\}$$

$$\{P_s\} = -12\lambda^2 p_o/S \quad \left\{\begin{array}{c} .25 \\ .50 \\ .70 \\ .25 \\ .50 \\ .75 \end{array}\right\}$$

in which

$p_o$ = the load intensity at the plate edge (Fig. 2.7);

$\lambda$ = the finite difference mesh width;

= $a/4$.

The deflection ordinates obtained as:

$$\left\{\begin{array}{c} w_{b1} \\ w_{b2} \\ w_{b3} \\ w_{b4} \\ w_{b5} \\ w_{b6} \end{array}\right\} = p_o a^4/147456. \ D \quad \left\{\begin{array}{c} 128.4 \\ 197.7 \\ 156.2 \\ 179.2 \\ 275.9 \\ 216.3 \end{array}\right\}$$

$$\left\{\begin{array}{c} w_{s1} \\ w_{s2} \\ w_{s3} \\ w_{s4} \\ w_{s5} \\ w_{s6} \end{array}\right\} = 3p_o a^2/4 \ S \quad \left\{\begin{array}{c} 0.020 \\ 0.035 \\ 0.033 \\ 0.027 \\ 0.045 \\ 0.043 \end{array}\right\}$$

## 2.7.3  PARTIALLY LOADED PLATE, Fig. 2.8

A simply supported square plate subjected to a partial load applied at its center is shown in Fig. 2.8. In the same figure, the finite difference mesh and the boundary conditions are shown. The coefficient matrices and load vectors are:

|  | $w_{b1}$ | $w_{b2}$ | $w_{b3}$ | $w_{b4}$ | $w_{b5}$ | $w_{b6}$ | $w_{b7}$ | $w_{b8}$ | $w_{b9}$ | $w_{b10}$ |  |
|---|---|---|---|---|---|---|---|---|---|---|---|
|  | 18. | -16. | 2. | 0. | 2. | 0. | 0. | 0. | 0. | 0. | 1 |
|  | -8. | 21. | -8. | 1. | -8. | 3. | 0. | 0. | 0. | 0. | 2 |
|  | 1. | -8. | 20. | -8. | 2. | -8. | 2. | 1. | 0. | 0. | 3 |
|  | 0. | 2. | -16. | 19. | 0. | 4. | -8. | 0. | 1. | 0. | 4 |
| [A] = | 2. | -16. | 4. | 0. | 20. | -16. | 2. | 2. | 0. | 0. | 5 |
|  | 0. | 3. | -8. | 2. | -8. | 23. | -8. | -8. | 3 | 0. | 6 |
|  | 0. | 0. | 4. | -8. | 2. | -16. | 20. | 4. | -8. | 1. | 7 |
|  | 0. | 0. | 2. | 0. | 2. | -16. | 4. | 22. | -16. | 2. | 8 |
|  | 0. | 0. | 0. | 1. | 0. | 6. | -8. | -16. | 25. | -8. | 9 |
|  | 0. | 0. | 0. | 0. | 0. | 0. | 4. | 8. | -32. | 20. | 10 |

|  | $w_{s1}$ | $w_{s2}$ | $w_{s3}$ | $w_{s4}$ | $w_{s5}$ | $w_{s6}$ | $w_{s7}$ | $w_{s8}$ | $w_{s9}$ | $w_{s10}$ |  |
|---|---|---|---|---|---|---|---|---|---|---|---|
|  | -58. | 32. | -2. | 0. | 0. | 0. | 0. | 0. | 0. | 0. | 1 |
|  | 16. | -59. | 16. | -1. | 16. | -1. | 0. | 0. | 0. | 0. | 2 |
|  | -1. | 16. | -60. | 16. | 0. | 16. | 0. | -1. | 0. | 0. | 3 |
|  | 0. | -2. | 32. | -59. | 0. | 0. | 16. | 0. | -1. | 0. | 4 |
| [B] = | 0. | 32. | 0. | 0. | -60. | 32. | -2. | 0. | 0. | 0. | 5 |
|  | 0. | -1. | 16. | 0. | 16. | -61. | 16. | 16. | -1. | 0. | 6 |
|  | 0. | 0. | 0. | 16. | -2. | 32. | -60. | 0. | 16. | -1. | 7 |
|  | 0. | 0. | -2. | 0. | 0. | 32. | 0. | -62. | 32. | 0. | 8 |
|  | 0. | 0. | 0. | -1. | 0. | -2. | 16. | 32. | -61. | 16. | 9 |
|  | 0. | 0. | 0. | 0. | 0. | 0. | -4. | 0. | 64. | -60. | 10 |

$$\{P_b\} \;=\; p_o\lambda^4/D \quad \left\{\begin{array}{c} 0.0 \\ 0.0 \\ 0.0 \\ 0.0 \\ 0.0 \\ \vdots \end{array}\right\}$$

$$\begin{Bmatrix} 0.0 \\ 0.0 \\ 0.5 \\ 0.5 \\ 1.0 \end{Bmatrix}$$

$$\{P_s\} = -12p_o\lambda^2/S \begin{Bmatrix} 0.0 \\ 0.0 \\ 0.0 \\ 0.0 \\ 0.0 \\ 0.0 \\ 0.0 \\ 0.5 \\ 0.5 \\ 1.0 \end{Bmatrix}$$

in which

$p_o$ = the partial load intensity;

$\lambda$ = the finite difference mesh width;

= $a/8$.

The deflection ordinates obtained as:

$$\{w_b\} = p_o a^2/4096. \ D \begin{Bmatrix} 0.399 \\ 0.756 \\ 1.016 \\ 1.112 \\ 1.443 \\ 1.957 \\ 2.148 \\ 2.695 \\ 2.975 \\ 3.302 \end{Bmatrix}$$

$$\{w_s\} = 3p_o a^2/16 \ S \begin{Bmatrix} 0.007 \\ 0.014 \\ 0.020 \\ 0.022 \\ 0.029 \end{Bmatrix}$$

$$\begin{Bmatrix} 0.044 \\ 0.050 \\ 0.074 \\ 0.086 \\ 0.105 \end{Bmatrix}$$

## 2.8   REFERENCES

[1]   Abel, J.F., and Popov, E.P., "Static and Dynamic Finite Element Analysis of Sandwich Structures", Proc. of the Conference on Matrix Methods in Structural Mechanics, TR-68-150, 1968, Air Force Flight Dynamics Lab., Wright-Patterson Air Force Base, Ohio.

[2]   Ahmed, K.N., "Mixed Finite-Difference Scheme for Analysis of Simply Supported Thick Plates", Computers and Structures, Vol. 3, 1973.

[3]   Allen, H.G., Analysis and Design of Structural Sandwich Panels, First edition, Pergamon Press Ltd., 1969.

[4]   Chan, H.C., and Cheung, Y.K., "Static and Dynamic Analysis of Multi-Layered Sandwich Plates", Int. J. Mech. Sci., Vol. 14, 1972.

[5]   Chong, S., "On the Theory of Bending of Sandwich Plates", Proc. of the 4th U.S. National Congress of Applied Mechanics (American Society of Mechanical Engineers), New York, 1962.

[6]   Cook, R.D., Concepts and Applications of Finite Element Analysis - A Treatment of the Finite Element Method as used for the Analysis of Displacement, Strain, and Stress, John Wiley and Sons, Inc., 1974.

[7]   Ericksen, W.S., "Effects of Shear Deformation in the Core of a Flat Rectangular Sandwich Panel", U.S. Forest Products Laboratory, Report No. 1583-C, December, 1950.

[8]   Folie, G.M., "Bending of Clamped Orthotropic Sandwich Plates", J. Eng. Mech. Div., ASCE, Vol. 96, No. EM3, June, 1970.

[9]   Ha, H.K., "Analysis of Three-Dimensional Orthotropic Sandwich Plate Structures by Finite Element Method", D. Eng. Thesis, Sir George Williams University, Montreal, Canada, 1972.

[10] Ha, K.H., and Fazio, P., "Flexural Behavior of Sandwich Floor Assembly", Building and Environment, Vol. 13, No. 1, 1978.

[11] Jaeger, L.G., Elementary Theory of Elastic Plates, Pergamon Press, 1964.

[12] Khatua, T.P., and Cheung, U.K., "Triangular Element for Multi-Layer Sandwich Plates", J. Eng. Mech. Div., ASCE, Vol. 98, No. EM5, October, 1972.

[13] Khatua, T.P., and Cheung, Y.K., "Bending and Vibration of Multi-Layer Sandwich Beams and Plates", International Journal for Numerical Methods in Engineering, Vol. 6, 1973.

[14] Kolar, V., and Nemec, I., "The Efficient Finite Element Analysis of Rectangular and Skew Laminated Plates", International Journal for Numerical Methods in Engineering, Vo. 7, 1973.

[15] March, H.W., "Effect of Shear Deformation in the Core of a Flat Retangular Sandwich Panel", U.S. Forest Products Laboratory, Report No. 1583, May, 1948.

[16] Mall, S.T., Tong, P., and Pian, T.H.H., "Finite Element Solutions for Laminated Thick Plates", J. Composite Materials, Vol. 6, April, 1972.

[17] Mawenya, A.S., and Davies, J.D., "Finite Element Bending Analysis of Multi-Layer Plates", Internation Journal for Numerical Methods in Engineering, Vol. 8, 1974.

[18] Monforton, G.R., and Schmit, L.A., "Finite Element Analysis of Sandwich Plates and Cylindrical Shells with Laminated Faces", Proc. of the Conference on Matrix Methods in Structural Mechanics, TR-68-150,, 1968, Air Force Flight Dynamics Lab., Wright-Patterson Air Force Base, Ohio.

[19] Monforton, G.R., and Michail, M.G., "Finite Element Analysis of Skew Sandwich Plates", Proc. of the American Society of Civil Engineers, Vol. 98, No. EM3, June, 1972.

[20] Monforton, G.R., and Ibrahim, I.M., "Analysis of Sandwich Plates with Unbalanced Cross-Ply Faces", Int. J. Mech. Sci., Vol. 17, 1975.

[21] Plantema, F.J., Sanwich Construction - The Bending and Buckling of Sandwich Beams, Plates and Shells, John Wiley and Sons, Inc., 1966.

[22] Ray, D.P., and Sinha, P.K., "On the Flexural Behavior of Orthotropic Sandwich Plates", Building Science, Vol. 8, 1973.

[23] Reissner, E., "On the Theory of Bending of Elastic Plates", J. Math. Phys., Vol. 23, 1942.

[24] Reissner, E., "The Effect of Transverse Shear Deformation on the Bending of Elastic Plates", J. Appl. Mech., Vol. 67, 1945.

[25] Reissner, E., "On Bending of Elastic Plates", Quarterly of Applied Mathematics, Vol. 5, 1947.

[26] Reissner, E., "Finite Deflection of Sandwich Plate", J. Aero. Sci., Vol. 15, 1948.

[27] Spilker, R.L., "Alternate Hybrid-Stress Elements for Analysis of Multi-Layer Compostie Plates", J. Composite Materials, Vol. 11, January, 1977.

[28] Szilard, R., Theory and Analysis of Plates - Classical and Numerical Methods, Prentice-Hall, Inc., 1974.

[29] Timoshenko, S., and Krieger, S., Theory of Plates and Shells, Second edition, McGraw-Hill Book Company, 1959.

[30] Ueng, C.E.S., and Lin, U.J., "On Bending of Orthotropic Sandwich Plates", J. American Inst. of Aeronautics and Astronautics, Vol. 4, No. 12, December, 1966.

[31] Yen, K.I., Gunturkun, S., and Pohle, F.V., "Deflections of a Simply Supported Rectangular Sandwich Plate Subjected to Transverse Loads", NACA, TN2581, December, 1951.

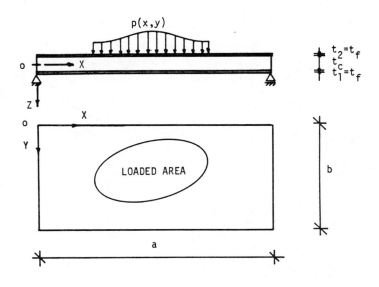

Fig. 2.1    COMPOSITE PLATE

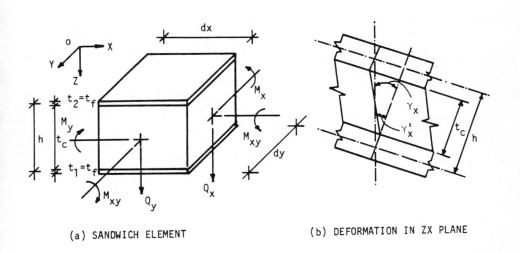

(a) SANDWICH ELEMENT

(b) DEFORMATION IN ZX PLANE

Fig. 2.2    SIGN CONVENTION AND SHEAR DEFORMATION OF A
COMPOSITE ELEMENT

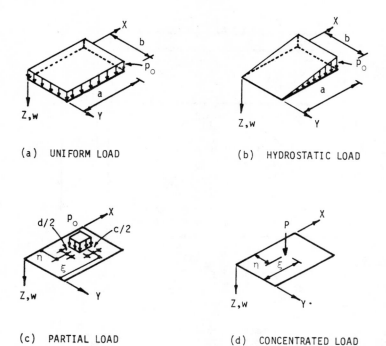

(a) UNIFORM LOAD

(b) HYDROSTATIC LOAD

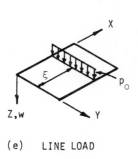

(c) PARTIAL LOAD

(d) CONCENTRATED LOAD

(e) LINE LOAD

Fig. 2.3  LOAD TYPES ON SIMPLY SUPPORTED COMPOSITE PLATE

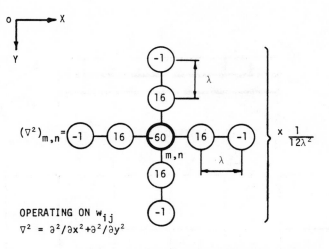

OPERATING ON $w_{ij}$

$\nabla^2 = \partial^2/\partial x^2 + \partial^2/\partial y^2$

Fig. 2.4 - STENCIL FOR HIGHER APPROXIMATION METHOD

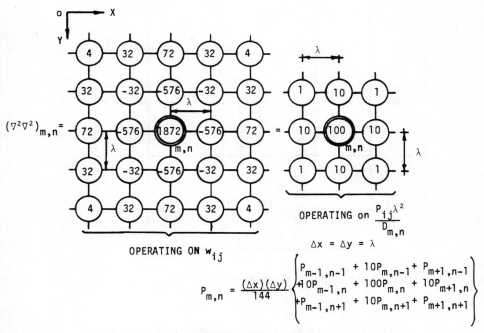

OPERATING ON $w_{ij}$

OPERATING on $\dfrac{P_{ij}\lambda^2}{D_{m,n}}$

$\Delta x = \Delta y = \lambda$

$$P_{m,n} = \frac{(\Delta x)(\Delta y)}{144}\left\{\begin{array}{l} P_{m-1,n-1} + 10P_{m,n-1} + P_{m+1,n-1} \\ +10P_{m-1,n} + 100P_{m,n} + 10P_{m+1,n} \\ +P_{m-1,n+1} + 10P_{m,n+1} + P_{m+1,n+1} \end{array}\right\}$$

Fig. 2.5 - STENCIL FOR FUNICULAR POLYGON METHOD

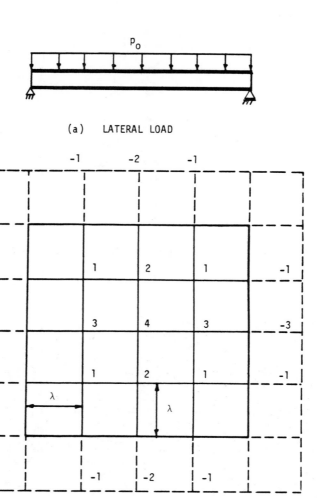

(a)   LATERAL LOAD

(b)   NUMBERING OF MESH POINTS

Fig. 2.6   UNIFORMLY LOADED SQUARE COMPOSITE PLATE

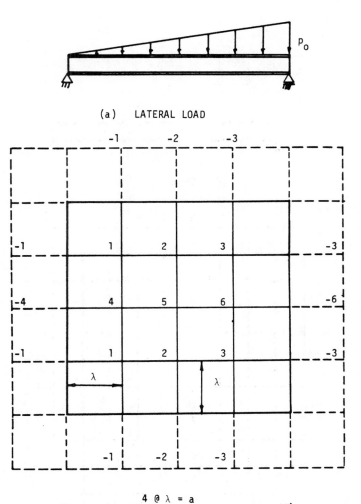

(a)  LATERAL LOAD

(b)  NUMBERING OF MESH POINTS

Fig. 2.7  SIMPLY SUPPORTED COMPOSITE PLATE SUBJECTED TO
HYDROSTATIC PRESSURE

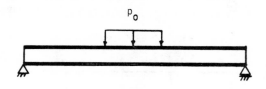

(a)  LATERAL LOAD

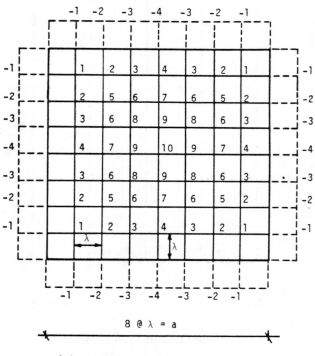

(b) NUMBERING OF MESH POINTS

Fig. 2.8    PARTIALLY LOADED COMPOSITE PLATE

58

# Continuous Composite Plates

## 3.1 INTRODUCTION

The problem of continuous composite plates is of considerable practical interest, since a floor or roof is usually subdivided by its supports into several panels not all of them are simply supported. In this chapter, analytic solutions for the deflection, shears, and moments in continuous composite plates subjected to lateral loads are developed. The plate considered is continuous over one intermediate support, and its panels have the same width while their spans are not necessarily equal or subjected to the same loading type.

This system is externally as well as internally statically indeterminate, and the available methods of analysis fall into two distinct categories: the force and deformation methods [4]. In either approach, the plate is divided into simple-span panels and the analysis is based on: (i) the equilibrium conditions of the individual spans and (ii) the compatibility of displacements or forces at the common support. In accordance with the force method, which is adopted in the present original analysis, the redundant distributed moment is first eliminated by introducing a fictitious hinge at the intermediate support as shown in Fig. 3.1(b). A means to determine this moment is provided by the compatibility condition which ensures the continuity of deformations of the adjacent panels. Once the redundant moment has been found, the deflections, shears, and moments at any point on either panel can be determined readily by combining the solutions of laterally loaded, simply supported composite plates established in the previous chapter with the effect of the edge moment as developed here. This general procedure is illustrated by the detailed development of the analytical solutions for two particular loading types: uniformly distributed and centrally concentrated loads acting on either panel. The application of the finite difference method to solve continuous composite plates is also presented.

Continuous sandwich beams have been analysed by Chong [1], Fazio and Salahuddin [2]. In the former work, experimental and theoretical analyses were conducted to investigate continuous beams composed of two equal spans and subjected to a uniformly distributed load throughout. A fourth order governing differential equation for the deflection was developed based on the displacement method, and for which a solution was found by superimposing the particular solution upon the homogeneous one. Fazio and Salahuddin [2] have studied the effect of settlement of a support of continuous sandwich beams on their redundant moments and shears. The analysis is based on the three-moment equation and considering the settlement of one support only, however, the combined effect of more than one support settlement can be treated by applying the principle of superposition. The theory was substantiated by experimental results obtained from tests on sandwich beams continuous over five supports and subjected to concentrated loads at their mid spans.

## 3.2 RECTANGULAR COMPOSITE PLATES SUBJECTED TO MOMENTS DISTRIBUTED ALONG THE EDGES

The analytic procedure for solving composite plates under moments distributed along the edges will be implemented using laminar composite plates. The number of laminae used here is three. Consider now a rectangular composite plate subjected to moments distributed along the edges $y = \pm b/2$, as shown in Fig. 3.2. The governing differential equation for the deflection in this case is obtained from Equ. (2.10) as:

$$\nabla\nabla w = 0 \tag{3.1}$$

A solution for the plate deflection, w, must satisy Equ. (3.1) as well as the following boundary conditions:

at  $x = 0, a$    $w = 0,$    $M_x = 0,$    and $\gamma_y = 0$ $\tag{3.2}$

at  $y = \pm b/2$    $w = 0$

and  $M_y = M_1$    for    $y = b/2$

or    $M_y = M_2$    for    $y = -b/2$ $\tag{3.3}$

in which

$M_1$ and $M_2$    =   the distributed moments along the edges $y = \pm b/2$, respectively.
a, b            =   plate dimensions in x- and y- directions, respectively.

A solution for Equ. (3.1) is taken in the form of the series [4, 5]

$$w = \sum_{m=1}^{\infty} Y_m(y)\, \sin \beta_m x \tag{3.4}$$

in which

$\beta_m = m\pi/a$

$\gamma$ is a function defined by:

$$\gamma_m(y) = A_m\sinh \beta_m y + B_m\cosh \beta_m y + C_m\beta_m y \sinh \beta_m y + D_m\beta_m y \cosh \beta_m y \qquad (3.5)$$

where $A_m$, $B_m$, $C_m$, and $D_m$ are unknown coefficients to be determined to satisfy the boundary conditions in Equs. (3.2) and (3.3).

To simplify the following analysis, three particular cases of moment distributions will be considered: (i) the symmetrical case in which $(M_y)_{y=b/2} = (M_y)_{y=-b/2}$ as shown in Fig. 3.3(a); (ii) the antisymmetrical case in which $(M_y)_{y=b/2} = -(M_y)_{y=-b/2}$ as shown in Fig. 3.3(b); and (iii) the antisymmetrical case in which $-(M_y)_{y-b/2} = (M_y)_{y=-b/2}$. The general case (fig. 3.2) can be obtained from these cases by applying the principle of superposition.

In the symmetrical case $\gamma_m$ must be an even function [3, 5]. Thus in Equ. (3.5), $A_m = D_m = 0$ and consequently

$$w_{sy} = \sum_{m=1}^{\infty} (B_m \cosh \beta_m y + C_m \beta_m y \sinh \beta_m y) \sin \beta_m x$$

in which

sy = subscript denotes the symmetrical case in Fig. 3.3(a).

By satisfying w = 0 in Equ. (3.3), this expression becomes:

$$w_{sy} = \sum_{m=1}^{\infty} C_m (\beta_m y \sinh \beta_m y - \alpha_m \tanh \alpha_m \cosh \beta_m y) \sin \beta_m x \qquad (3.6)$$

in which

$\alpha_m = m\pi/2R$;

R = the plate aspect ration = a/b.

Expressions for the shear force can be determined from Equ. (3.6) and are taken as:

$$Q_{xsy} = \sum_{m=1}^{\infty} (E_m \cosh \beta_m y + H_m \beta_m y \sinh \beta_m y) \cos \beta_m x$$

$$(3.7)$$

$$Q_{ysy} = \sum_{m=1}^{\infty} (F_m \sinh \beta_m y + P_m \beta_m y \cosh \beta_m y) \sin \beta_m x$$

in which
$E_m, H_m, F_m, P_m,$ = unknown coefficients.

By substituting Equs. (3.6) and (3.7) in the governing differential equations for shears, i.e. Equs. (2.11) and (2.12), it is found that:

$H_m = P_m = 0$
$E_m = F_m = K_m C_m$

in which
$K_m = -2D\beta_m^3$

Hence, Equ. (3.7) becomes:

$$Q_{xsy} = \sum_{m=1}^{\infty} K_m C_m \cosh \beta_m y \cos \beta_m x \qquad (3.8)$$

$$Q_{ysy} = \sum_{m=1}^{\infty} K_m C_m \sinh \beta_m y \sin \beta_m x \qquad (3.9)$$

The secondary boundary condition in Equ. (3.3) can be used now to determine the coefficient $C_m$. Representing the moment distribution along the edges $y = \pm b/2$ by a trigonometric series, as:

$$M_{sy} = \sum_{m=1}^{\infty} M_{msy} \sin \beta_m x \qquad (3.10)$$

in which
$M_{sy}$ = the symmetrical moment distribution (Fig. 3.3(a)).
$M_{msy}$ = the coefficient of Fourier expansion of the symmetrical moment, $M_{sy}$.

An expression for the moment $M_y$ is obtained from Equs. (2.2), (3.6), (3.8), and (3.9). This expression and Equ. (3.10) together with the second condition in Equ. (3.3) result in

$$C_m = \frac{-b^2 \, M_{msy}}{8 \, \alpha_m^2 \, D \, \xi_m \, \cosh \alpha_m} \tag{3.11}$$

in which

$$\xi_m = 1 + \frac{4D}{b^2 \, S} \, (1 - \nu) \, \alpha_m^2$$

Having obtained an expression for $C_m$, the deflection and shears of a plate subjected to symmetrical moment can be calculated readily from Equs. (3.6), (3.8), and (3.9). By substituting Equ. (3.11) in (3.6), the deflection surface is obtained as:

$$w_{sy} = \sum_{m=1}^{\infty} \frac{b^2 \, M_{msy}}{8 \, \alpha_m^2 \, D \, \xi_m \, \cosh \alpha_m} \, (\alpha_m \tanh \alpha_m \cosh \beta_m y$$

$$- \beta_m y \sinh \beta_m y) \sin \beta_m x \tag{3.12}$$

In the particular case of uniform distribution of the edge moment

$$M_{msy} = 4M_0/m\pi \quad m = 1, 3, 5, \ldots$$

in which
$M_o$ = the intensity of the uniform moment.

and consequently the deflection along the axis of symmetry $y = 0$ is

$$w_{sy} = \frac{2 \, M_o \, a^2}{\pi^3 \, D} \sum_{m=1,3,\ldots}^{\infty} \frac{1}{m^3 \, \xi_m} \, \frac{\alpha_m \tanh \alpha_m}{\cosh \alpha_m} \, \sin \beta_m x \tag{3.13}$$

When a is very large in comparison to b: $\tanh \alpha_m = \alpha_m$ and $\cosh \alpha_m = \xi_m = 1$, and the previous expression becomes:

$$w_{sy} = M_o b^2/8D$$

which is the deflection at the middle of a composite strip of length b, as to be expected.

For the antisymmetrical case shown in Fig. 3.3(b), the deflection surface must be an odd function of y [3, 5]. Hence $B_m = C_m = 0$ in Equ. (3.5), and Equ. (3.4) becomes

$$w_{asy} = \sum_{m=1}^{\infty} (A_m \sinh \beta_m y + D_m \beta_m y \cosh \beta_m y) \sin \beta_m x$$

in which
     asy = subscript denoting the antisymmetrical case in Fig. 3.3(b).

By satisfying w = 0 in Equ. (3.3), it is found that:

$$w_{asy} = \sum_{m=1}^{\infty} A_m (\sinh \beta_m y - \frac{\beta_m y}{\alpha_m} \tanh \alpha_m \cosh \beta_m y) \sin \beta_m x \quad (3.14)$$

The shear forces in this case take the form

$$Q_{xasy} = \sum_{m=1}^{\infty} (E'_m \sinh \beta_m y + H'_m \beta_m y \cosh \beta_m y) \cos \beta_m x$$

$$\left.\begin{array}{c} \\ \\ \end{array}\right\} \quad (3.15)$$

$$Q_{yasy} = \sum_{m=1}^{\infty} (F'_m \cosh \beta_m y + P'_m \beta_m y \sinh \beta_m y) \sin \beta_m x$$

in which
     $E_m', H_m', F_m', P_m'$ = unknown coefficients.

From Equ. (3.15) together with the governing differential equations for shears, i.e. Equs. (2.11) and (2.12), it is found that:

$$H_m' = P_m' = 0$$

and

$$E_m' = F_m' = K_m'\dot{A}_m$$

in which

$$K_m' \quad = \quad 2\ D\ \beta_m^3\ \frac{\tanh\ \alpha_m}{\alpha_m}$$

Consequently, Equ. (3.15) becomes:

$$Q_{xasy} \quad = \quad \sum_{m=1}^{\infty}\ K_m'\ A_m\ \sinh\ \beta_m y\ \cos\ \beta_m x \qquad (3.16)$$

$$Q_{yasy} \quad = \quad \sum_{m=1}^{\infty}\ K_m'\ A_m\ \cosh\ \beta_m y\ \sin\ \beta_m x \qquad (3.17)$$

Having obtained Equs. (3.14), (3.16), and (3.17), the coefficient $A_m$ can be easily determined. Representing the antisymmetrical moment distribution along the edges y = ±b/2 by a trigonometric series:

$$M_{asy} \quad = \quad \sum_{m=1}^{\infty}\ M_{masy}\ \sin\ \beta_m x \qquad (3.18)$$

in which
  $M_{asy}$ = the antisymmetrical moment distribution, Fig. 3.3(b)
  $M_{masy}$ = the coefficient of Fourier expansion of the edges antisymmetrical moment, $M_{asy}$.

From which and Equs. (2.2), (3.14), (3.16), and (3.17), together with the second boundary condition in Equ. (3.3), it follows that:

$$A_m \quad = \quad \frac{b^2\ M_{masy}}{8\ \alpha_m\ D\ \xi_m\ \cosh\ \alpha_m}$$

For the antisymmetrical case in which $(M_y)_{y=b/2} = (M_y)_{y=-b/2}$, the deflection surface and the shear forces take the same forms as in Equs. (3.14), (3.16), and (3.17). The coefficient $A_m$ in this case is determined by following the same procedure and the expression obtained is

$$A_m' = \frac{-b^2 \, M_{masy}}{8 \, \alpha_m \, D \, \xi_m \, \cosh \, \alpha_m}$$

where $A_m'$ is the coefficient which replaces $A_m$ in Equs. (3.14), (3.16), and (3.17).

Finally, for the general case, represented by the second and third boundary conditions in Equ. (3.3), the deflection surface and shears can be obtained from the solutions developed for the symmetrical and antisymmetrical cases. For this purpose, the moment distribution $M_1$ and $M_2$ can be split into a symmetrical distribution $M_y'$ and an antisymmetrical distribution $M_y''$ as follows:

$$(M_y')_{y=b/2} \quad = \quad (M_y')_{y=-b/2} \quad = \quad [M_1(x) + M_2(x)]/2$$

$$(M_y'')_{y=b/2} \quad = \quad - (M_y'')_{y=-b/2} \quad = \quad [M_1(x) - M_2(x)]/2$$

By representing $M_y'$ and $M_y''$ by trigonometric series, the general solution can be easily determined. When a moment $M_c$ is distributed only along one edge (Fig. 3.4), either $M_1(x) = 0$ or $M_2(x) = 0$, and from the previous equations $M_{msy} = M_{masy} = M_m/2$ where $M_m$ is the coefficient of the trigonometric expansion of $M_c$:

$$M_c \quad = \quad \sum_{m=1}^{\infty} M_m \, \sin \, \beta_m x \qquad\qquad (3.19)$$

Solutions for two particular cases are presented next. When $M_c = (M_y)_{y=b'/2}$, as shown in Fig. 3.4(a), the deflection surface and shear forces are obtained as:

$$w^r = \sum_{m=1}^{\infty} \frac{b^{r2} M_m^r}{16 \, \alpha_m^{r2} \, D \, \xi_m^r} \left[ \frac{1}{\cosh \alpha_m^r} (\alpha_m^r \tanh \alpha_m^r \cosh \beta_m y \right.$$

$$- \beta_m y \sinh \beta_m y) + \frac{1}{\sinh \alpha_m^r} (\alpha_m^r \coth \alpha_m^r \sinh \beta_m y$$

$$\left. - \beta_m y \cosh \beta_m y) \right] \sin \beta_m x \tag{3.20}$$

$$Q_x^r = \sum_{m=1}^{\infty} \frac{\alpha_m^r M_m^r}{b^r \, \xi_m^r} \left( \frac{\cosh \beta_m y}{\cosh \alpha_m^r} + \frac{\sinh \beta_m y}{\sinh \alpha_m^r} \right) \cos \beta_m x \tag{3.21}$$

$$Q_y^r = \sum_{m=1}^{\infty} \frac{\alpha_m^r M_m^r}{b^r \, \xi_m^r} \left( \frac{\sinh \beta_m y}{\cosh \alpha_m^r} + \frac{\cosh \beta_m y}{\sinh \alpha_m^r} \right) \sin \beta_m x \tag{3.22}$$

in which

$b^r$ = the plate dimension parallel to y- axis;

$R^r$ = the plate aspect ratio

= $a/b^r$;

$M_m^r$ = the coefficient of the series in Equ. (3.19);

$\alpha_m^r$ = $m\pi/2R^r$;

$\xi_m^r$ = $1 + (4D/b^{r2}S)(1 - \upsilon)\alpha_m^{r2}$

The moments of the plate in this case are determined from Equs. (3.20), (3.21), and (3.22) in conjunction with Equs. (2.1), (2.2), and (2.3). In abreviated forms, the results are:

$$M_x^r = - D \left( \frac{\partial \theta_x^r}{\partial x} + \upsilon \frac{\partial \theta_y^r}{\partial y} \right)$$

$$M_y^r = - D \left( \frac{\partial \theta_y^r}{\partial y} + \upsilon \frac{\partial \theta_x^r}{\partial x} \right) \tag{3.23}$$

$$M_{xy}^r = D_{xy} \left( \frac{\partial \theta_x^r}{\partial y} + \frac{\partial \theta_y^r}{\partial x} \right)$$

in which

$$\theta_x^r = \frac{\partial w^r}{\partial x} - \frac{Q_x^r}{S}$$

$$= \sum_{m=1}^{\infty} \frac{b^r M_m^r}{8\, \alpha_m^r\, D\, \xi_m^r} \left[ \frac{1}{\cosh \alpha_m^r} \left( -\beta_m y \sinh \beta_m y \right.\right.$$

$$\left. + \alpha_m^r \tanh \alpha_m^r \cosh \beta_m y - 8\, \frac{\alpha_m^{r2}\, D}{b^{r2}\, S} \cosh \beta_m y \right)$$

$$+ \frac{1}{\sinh \alpha_m^r} \left( \alpha_m^r \coth \alpha_m^r \sinh \beta_m y - \beta_m y \cosh \beta_m y \right.$$

$$\left.\left. - 8\, \frac{\alpha_m^{r2}\, D}{b^{r2}\, S} \sinh \beta_m y \right) \right] \cos \beta_m x$$

$$\theta_y^r = \frac{\partial w^r}{\partial y} - \frac{Q_y^r}{S}$$

$$= \sum_{m=1}^{\infty} \frac{b^r M_m^r}{8\, \alpha_m^r\, D\, \xi_m^r} \left[ \frac{1}{\cosh \alpha_m^r} \left( -\sinh \beta_m y - \beta_m y \cosh \beta_m y \right.\right.$$

$$\left. + \alpha_m^r \tanh \alpha_m^r \sinh \beta_m y - 8\, \frac{\alpha_m^{r2}\, D}{b^{r2}\, S} \sinh \beta_m y \right)$$

$$+ \frac{1}{\sinh \alpha_m^r} \left( \alpha_m^r \coth \alpha_m^r \cosh \beta_m y - \cosh \beta_m y \right.$$

$$\left.\left. - \beta_m y \sinh \beta_m y - 8\, \frac{\alpha_m^{r2}\, D}{b^{r2}\, S} \cosh \beta_m y \right) \right] \sin \beta_m x \qquad (3.24)$$

When the moment $M_c$ is acting on the edge $y = -b^{\ell}/2$ as shown in Fig. 3.4(b), the deflection and shears are obtained as:

$$w^{\ell} = \sum_{m=1}^{\infty} \frac{b^{\ell 2} M_m^{\ell}}{16 \alpha_m^{\ell 2} D \xi_m^{\ell}} [\frac{1}{\cosh \alpha_m^{\ell}} (- \beta_m y \sinh \beta_m y$$

$$+ \alpha_m^{\ell} \tanh \alpha_m^{\ell} \cosh \beta_m y) - \frac{1}{\sinh \alpha_m^{\ell}} (\alpha_m^{\ell} \coth \alpha_m^{\ell} \sinh \beta_m y$$

$$- \beta_m y \cosh \beta_m y)] \sin \beta_m x \tag{3.25}$$

$$Q_x^{\ell} = \sum_{m=1}^{\infty} \frac{\alpha_m^{\ell} M_m^{\ell}}{b^{\ell} \xi_m^{\ell}} (\frac{\cosh \beta_m y}{\cosh \alpha_m^{\ell}} - \frac{\sinh \beta_m y}{\sinh \alpha_m^{\ell}}) \cos \beta_m x \tag{3.26}$$

$$Q_y^{\ell} = \sum_{m=1}^{\infty} \frac{\alpha_m^{\ell} M_m^{\ell}}{b^{\ell} \xi_m^{\ell}} (\frac{\sinh \beta_m y}{\cosh \alpha_m^{\ell}} - \frac{\cosh \beta_m y}{\sinh \alpha_m^{\ell}}) \sin \beta_m x \tag{3.27}$$

in which

$b^{\ell}$ = the plate dimension parallel to y- axis

$\alpha_m^{\ell}$ = $\dfrac{m\pi}{2 R^{\ell}}$

$R^{\ell}$ = the plate aspect ratio

= $a/b^{\ell}$

$M_m^\ell$  =  the coefficient of the series in Equ. (3.19)

$\xi_m^\ell$  =  $1 + \dfrac{4\,D}{b^{\ell 2}\,S}\,(1-\nu)\,\alpha_m^{\ell 2}$

The moments in this case can be obtained from Equ. (3.23), noting that $\theta_x{}''$ and $\theta_y{}''$ should replace $\theta_x{}^r$ and $\theta_y{}^r$, respectively, where

$$
\theta_x^\ell = \sum_{m=1}^{\infty} \frac{b^\ell\,M_m^\ell}{8\,\alpha_m^\ell\,D\,\xi_m^\ell}\,[\frac{1}{\cosh\alpha_m^\ell}\,(-\beta_m y\,\sinh\beta_m y
$$

$$
+\,\alpha_m^\ell\,\tanh\alpha_m^\ell\,\cosh\beta_m y\,-\,8\,\frac{\alpha_m^{\ell 2}\,D}{b^{\ell 2}\,S}\,\cosh\beta_m y)
$$

$$
-\,\frac{1}{\sinh\alpha_m^\ell}\,(\alpha_m^\ell\,\coth\alpha_m^\ell\,\sinh\beta_m y\,-\,\beta_m y\,\cosh\beta_m y
$$

$$
-\,8\,\frac{\alpha_m^{\ell 2}\,D}{b^{\ell 2}\,S}\,\sinh\beta_m y)]\,\cos\beta_m x
$$

$$
\theta_y^\ell = \sum_{m=1}^{\infty} \frac{b^\ell\,M_m^\ell}{8\,\alpha_m^\ell\,D\,\xi_m^\ell}\,[\frac{1}{\cosh\alpha_m^\ell}\,(-\sinh\beta_m y\,-\,\beta_m y\,\cosh\beta_m y
$$

$$
+\,\alpha_m^\ell\,\tanh\alpha_m^\ell\,\sinh\beta_m y\,-\,8\,\frac{\alpha_m^{\ell 2}\,D}{b^{\ell 2}\,S}\,\sinh\beta_m y)
$$

$$
-\,\frac{1}{\sinh\alpha_m^\ell}\,(\alpha_m^\ell\,\coth\alpha_m^\ell\,\cosh\beta_m y\,-\,\cosh\beta_m y
$$

$$
-\,\beta_m y\,\sinh\beta_m y\,-\,8\,\frac{\alpha_m^{\ell 2}\,D}{b^{\ell 2}\,S}\,\cosh\beta_m y)]\,\sin\beta_m x \qquad (3.28)
$$

## 3.3 CONTINUOUS COMPOSITE PLATES

A rectangular plate of width a and length $(b'' + b^r)$, supported along the edges and also along the intermediate line bb (Fig. 3.1(a)) forms a simply supported continuous plate over two spans [5]. Solutions for its deflected surface, shears, and moments can be obtained by combining those obtained for laterally loaded, simply supported plates (treated in chapter II), with those developed for plates subjected to distributed moments along their edges (section 3.2). This procedure requires the knowledge of the redundant moment which is of such a magnitude as to ensure compatibility of the deformations of the two adjacent panels at the common support.

Consider now the simply supported continuous laminar plate which is subjected to a lateral load on both of its spans as shown in Fig. 3.1(a). The slope of the bending deflection surface along the edge y = b of the laterally loaded, simply supported right span (Fig. 3.5(a)) is obtained form Equ. (2.14) as:

$$\left(\frac{\partial w_b^{r'}}{\partial y^r}\right)_{y^r=b^r} = \sum_{m=1}^{\infty} C_m^r \sin \beta_m x \qquad (3.29)$$

in which

$w_b^{r'}$ = the bending component of the total deflection of the

   right span

$X^r$, $Y^r$ = the coordinate axes for the right panel (Fig. 3.5 (a))

$$C_m^r = \sum_{n=1}^{\infty} \frac{n\pi}{Db^r} \frac{(-1)^{(n+2)} P_{mn}^r}{[(m\pi/a)^2 + (n\pi/b^r)^2]^2}$$

where $P_{mn}{}^r$ is the coefficient of Fourier expansion for the applied load on the right panel. Let $M_c$ be the redundant moment distributed along the edge $y^r = b^r$ (Fig. 3.5(a)), and be represented by the sine series in Equ. (3.19). The deflection surface due to $M_c$

is represented by Equ. (3.20), and its bending slope $\theta_y^r$ along the edge $y = b^r/2$ (Fig. 3.5(b)) is determined from Equ. (3.24) as:

$$
\theta_y^{r''} = \sum_{m=1}^{\infty} \frac{b^r M_m}{8 \alpha_m^r D \xi_m^r} \left(- \tanh \alpha_m^r - 2 \alpha_m^r + \alpha_m^r \tanh^2 \alpha_m^r \right.
$$

$$
- 8 \alpha_m^{r2} \phi^r \tanh \alpha_m^r + \alpha_m^r \coth^2 \alpha_m^r - \coth \alpha_m^r
$$

$$
\left. - 8 \alpha_m^{r2} \phi^r \coth \alpha_m^r \right) \sin \beta_m x \tag{3.30}
$$

in which

$\phi^r$  =  the shear parameter of the right panel

$\quad$ = $\dfrac{D}{b^{r2} S}$

$\theta_y^{r''}$ = the slope of the bending deflection surface of the plate

$\quad$ due to the moment $M_c$ distributed along the edge $y^r = b^r/2$.

The slope of the bending deflection surface along the edge $y'' = 0$ of the laterally loaded simply supported left panel (Fig. 3.5(a)) is obtained from Equ. (2.14) as:

$$
\left( \frac{\partial w_b^{\ell'}}{\partial y^{\ell}} \right)_{y^{\ell}=0} = \sum_{m=1}^{\infty} C_m^{\ell} \sin \beta_m x \tag{3.31}
$$

in which

$$w_b^{\ell\,\prime} \quad = \quad \text{the same as } w_b^{r\,\prime} \text{ but for the left panel}$$

$$x^\ell, \; Y^\ell \quad = \quad \text{coordinate axes for the left panel (Fig. 3.5 (a))}$$

$$C_m^\ell \quad = \quad \sum_{n=1}^{\infty} \frac{n\pi}{Db^\ell} \frac{p_{mn}^\ell}{[(m\pi/a)^2 + (n\pi/b^\ell)^2]^2}$$

where $p_{mn}\text{"}$ is the coefficient of Fourier expansion of the load acting on the left panel. Due to the moment $M_c$ (Equ. 3.19) which is distributed along the edge $y\text{"} = 0$ (Fig. 3.5(a)), the deflection can be determined from Equ. (3.25) and its bending slope $\theta_{y\text{"}}$ along the edge $y\text{"} = -b\text{"}/2$ (Fig. 3.5(b)) is obtained form Equ. (3.28) as:

$$\theta_y^{\ell\,\prime\prime} \quad = \quad \sum_{m=1}^{\infty} \frac{b^\ell M_m}{8 \alpha_m^\ell D \xi_m^\ell} (\tanh \alpha_m^\ell + 2 \alpha_m^\ell - \alpha_m^\ell \tanh^2 \alpha_m^\ell$$

$$+ 8 \alpha_m^{\ell 2} \phi^\ell \tanh \alpha_m^\ell - \alpha_m^\ell \coth^2 \alpha_m^\ell$$

$$+ \coth \alpha_m^\ell + 8 \alpha_m^{\ell 2} \phi^\ell \coth \alpha_m^\ell) \sin \beta_m x \qquad (3.32)$$

From the condition of continuity it is concluded that the sum of the expressions in the right hand side of Equs. (3.29) and (3.30), which represents the bending slope of the right panel, of the continuous plate in Fig. 3.1(a), along the intermediate support must be equal to that in the left panel. Thus

$$\left(\frac{\partial w_b^{r\,\prime}}{\partial y^r}\right)_{y^r = b^r} + \theta_y^{r\,\prime\prime} \quad = \quad \left(\frac{\partial w_b^{\ell\,\prime}}{\partial y^\ell}\right)_{y^\ell = 0} + \theta_y^{\ell\,\prime\prime}$$

By substituting Equs. (3.29), (3.30), (3.31), and (3.32) in this equation, and by equating the coefficients of the sine terms in both sides it is found that:

$$M_m \quad = \quad \frac{\phi_m}{\psi_m} \qquad\qquad\qquad (3.33)$$

in which

$$\phi_m = C_m^r - C_m^\ell$$

$$\Psi_m = \frac{b^\ell}{8 D} \Psi_m' \qquad\qquad (3.34)$$

where

$$\Psi_m' = \frac{1}{\alpha_m^\ell \xi_m^\ell} [(\tanh \alpha_m^\ell + \coth \alpha_m^\ell)$$

$$(1 + 8 \alpha_m^{\ell 2} \phi^\ell) + \alpha_m^\ell (2 - \tanh^2\alpha_m^\ell - \coth^2\alpha_m^\ell)]$$

$$+ \frac{b^r}{b^\ell} \frac{1}{\alpha_m^r \xi_m^r} [(\tanh \alpha_m^r + \coth \alpha_m^r) (1 + 8 \alpha_m^{r2} \phi^r)$$

$$+ \alpha_m^r (2 - \tanh^2\alpha_m^r - \coth^2\alpha_m^r)]$$

Two particular loading types are now considered. First, when the applied load is uniformly distributed:

$$\phi_m = \frac{16 a^4}{\pi^5 D b^r} \phi_{mu}' \qquad\qquad (3.35)$$

in which

$$\phi_{mu}' = \sum_{n=1,3,\ldots}^\infty [\frac{(-1)^{(n+2)} p_o^r}{m(m^2 + n^2 R^{r2})^2} - \frac{b^r}{b^\ell} \frac{p_o^\ell}{m(m^2 + n^2 R^{\ell 2})^2}]$$

where $p_o^r$ and $p_o{''}$ are the intensities of the uniform loads acting on the right and left panels respectively. Substituting Equs. (3.34) and (3.35) in Equ. (3.33) gives

$$M_m = \frac{128 a^4}{\pi^5 b^r b^\ell} \frac{\phi_{mu}'}{\Psi_m'} \qquad\qquad (3.36)$$

The right side of this equation represents Fourier expansion coefficent of the redundant moment of a simply supported continuous composite plate (Fig. 3.1(a)) subjected to uniform loads on both panels. Knowing the coefficient $M_m$ from Equ. (3.36), the value of the redundant moment $M_c$, at any point along the common support can be obtained from Equ. (3.19). The value of this moment at $x = a/2$, that is at the middle of the width of the plate is obtained as

$$(M_c)_{\substack{y^r = b^r/2 \\ x^r = a/2}} = b^{r2} \sum_{m=1,3,\ldots}^{\infty} \frac{-128}{\pi^5} (-1)^{(m+1)/2}$$

$$\frac{R^{\ell 2} R^{r2}}{k} \frac{\phi'_{mu}}{\psi'_m} \qquad (3.37)$$

in which

$$k = \frac{b^r}{b^\ell} = \frac{R^\ell}{R^r}$$

Second, when the external load is a single concentrated laod at the center of each panel:

$$\phi_m = \frac{4 a^3}{\pi^3 D b^{r2}} \phi'_{mc} \qquad (3.38)$$

in which

$$\phi'_{mc} = \sum_{n=1}^{\infty} -n (-1)^{(m+n)/2} \left[ \frac{(-1)^{(n+2)} p^r}{(m^2 + n^2 R^{r2})^2} \right.$$

$$\left. - \frac{b^{r2}}{b^{\ell 2}} \frac{p^\ell}{(m^2 + n^2 R^{\ell 2})^2} \right]$$

where $p^r$ and $p''$ are the central concentrated loads on the right and left panels respectively.

The value of the redundant moment at $x = a/2$, in this case is obtained from Equs. (3.33), (3.34), (3.38), together with Equ. (3.19) as:

$$(M_c)_{\substack{y^r = b^r/2 \\ x^r = a/2}} = \sum_{m=1,3,\ldots}^{\infty} \frac{-32}{\pi^3} (-1)^{(m+1)/2} k R^{r3} \frac{\phi'_{mc}}{\psi'_m} \qquad (3.39)$$

## 3.4   PRACTICAL FORMULAS FOR COMPOSITE PLATES SUBJECTED TO EDGES MOMENT

To reduce the amount of work involved in the analysis of continuous composite plates, practical formulas are devised from which the deflection $w$, moments $M_x$ and $M_y$, and shear forces $Q_x$ and $Q_y$ can be determined. The exact, simple expressions developed contain two main factors. The first one reflects the material properties and the plate geometry, whereas the second reflects the aspect ratio of the panels, the ratio of the right to left spans, the loading type and the shear parameter which is defined as:

$$\text{shear parameter} = \phi^i$$

$$= \frac{D}{S \, b^{i2}}$$

in which

$$i = r \text{ or } \ell$$

where $r$ and „ denote the right and left panels respectively.

Numerical results are tabulated over a certain range for each of the parameters as follows:

    (i)    the ratio of the right to left spans, $k = b^r/b$„, varies from 0.25 to 1.
    (ii)   the aspect ratio of the right panel, $R^r = a/b^r$, varies from 1. to 5.
    (iii)  the shear parameter of the right panel, $\phi^r$, varies from 0.0 to 0.5.

Three loading types are considered: (i) a uniformly distributed moment along the edges $y = \pm b/2$ (Fig. 3.6), (ii) a uniformly distributed load (Fig. 3.6), and (iii) a central concentrated load (Fig. 3.6). To extend further the possible application of the present formulas, the uniform and concentrated loads are considered acting on either the right or left panel. The combined effect of more than one loading type can be obtained by applying the principle of superposition. For each loading type, the numerical values of the factors in the practical formulas are tabulated in Apendix B.

In these tables, the following notations are used

| | | |
|---|---|---|
| SPR = $\phi^r$ | = | the shear parameter for the right panel (Fig. 3.1) |
| SPL = $\phi$" | = | the shear parameter for the left panel |
| | = | $k^2 \cdot \phi^r$ |
| POR, PR | = | $p_o{}^r$, $p^r$, respectively. |
| POL, PL | = | $p_o$" , $p$" , respectively. |
| | | where $p_o{}^i$ and $p^i$ are the intensity of a uniformly distributed load and the value of a central concentrated load acting on panel i, respectively. |
| k | = | $b^r/b$" $= R$"$/R^r$ |
| | | where $b^i$ and $R^i$ are the span and aspect ratio of panel i. |
| POIS.R | = | Poisson's ratio of outer materials. |
| N, F | = | Denote the intermediate support and the one parallel to it, respectively. Thus, for example, $K_{Qyn}{}^r$ means that it is the factor of the shear force $Q_y$ at the middle of the intermediate support in the right panel (Fig. 3.6). Additional notations are shown in Fig. 3.6. |
| RR, RL | = | $R^r$, $R$", respectively. |
| KWR, KWL | = | $K_w{}^r$, $K_w$ ", respectively. |
| KMC | = | $K_{mc}$ |
| KQXR, KQXL | = | $K_{Qx}{}^r$, $K_{Qx}$", respectively. |
| KQYRN, KQYLN | = | $K_{Qyn}{}^r$, $K_{Qyn}$" , respectively. |
| KQYRF, KQYLF | = | $K_{Qyf}{}^r$, $K_{Qyf}$" , respectively. |
| KMXR, KMXL | = | $K_{mx}{}^r$, $K_{mx}$" , respectively. |
| KMYR, KMYL | = | $K_{my}{}^r$, $K_{my}$" , respectively. |

It should be emphasized that all the tabulated values as well as formulas presented in the following are the response of continuous plates due to only the redundant moment. The principle of superposition must be applied to determine the total effect of the applied loads.

### 3.4.1 COMPOSITE PLATE SUBJECTED TO UNIFORM MOMENTS AT y = ±b/2 (Fig. 3.6)

The central deflection in this case can be determined from Equ. (3.13) as:

$$w_{sy} = \frac{M_o a^2}{D} K_{wsy} \qquad (3.40)$$

in which

$$K_{wsy} = \sum_{m=1,3,\ldots}^{\infty} \frac{-(-1)^{(m+1)/2}}{\pi^2 R \xi_m m^2} \frac{\tanh \alpha_m}{\cosh \alpha_m}$$

The factor $K_{wsy}$ is evaluated for a wide range of aspect ratios and shear parameters and the results are shown in Fig. 3.7.

For the particular case when the plate is rigid in shear: $\xi_m = 1$, and Equ. (3.40) becomes:

$$w_{sy} = \frac{M_o b^2}{D} \sum_{m=1,3,\ldots}^{\infty} \frac{-(-1)^{(m+1)/2}}{\pi^2 \xi_m m^2} R \frac{\tanh \alpha_m}{\cosh \alpha_m}$$

Some numerical results are obtained from this expression and shown in Table B.1.

The bending moments $M_x$ and $M_y$ at the plate center, are obtained from Equs. (3.13), (3.16), (3.17) in conjunction with Equs. (2.1) and (2.2), as:

$$M_x = M_o K_{mxsy} \tag{3.41}$$

$$M_y = M_o K_{mysy} \tag{3.42}$$

in which

$$K_{mxsy} = \sum_{m=1,3,\ldots}^{\infty} \frac{-2 (-1)^{(m+1)/2} R}{\pi^2 \xi_m m^2 \cosh \alpha_m} [2 \alpha_m^2 (1-\nu) \tanh \alpha_m$$

$$+ 4 \alpha_m \nu + 16 \alpha_m^3 \phi (\nu-1)]$$

$$K_{mysy} = \sum_{m=1,3,\ldots}^{\infty} \frac{-2 (-1)^{(m+1)/2} R}{\pi^2 \xi_m m^2 \cosh \alpha_m} [4 \alpha_m + 2 \alpha_m^2 (\nu-1) \tanh \alpha_m$$

$$+ 16 \alpha_m^3 \phi (1-\nu)$$

The higher normal stress in the faces corresponds to the bending moment $M_y$, and consequently the numerical values of only $K_{mysy}$ are shown in Fig. 3.7.

When $\xi_m = 1$, some numerical results of $M_x$ and $M_y$ are shown in Table B.1. Observation of this table shows that the entire applied edge moment $M_o$ is transmitted by the transverse section at the middle of a strip whereas the bending moment $M_y$ at the center of a plate decreases, as compared with $M_o$, with decreasing ratio a/b. This is due to a damping effect of the edges x = o and x = a which are not exposed to couples [5].

The shear forces at the middle of the edges x = 0 and y = b/2 are determined from Equs. (3.8) and (3.9) as:

$$Q_{xsy} = \frac{4 \, M_o}{a} \, K_{Qxsy} \tag{3.43}$$

$$Q_{ysy} = \frac{4 \, M_o}{a} \, K_{Qysy} \tag{3.44}$$

in which

$$K_{Qxsy} = \sum_{m=1,3,\ldots}^{\infty} \frac{1}{\xi_m \, \cosh \alpha_m}$$

$$K_{Qysy} = \sum_{m=1,3,\ldots}^{\infty} \frac{- (-1)^{(m+1)/2} \, \tanh \alpha_m}{\xi_m}$$

The numerical values for only the factor $K_{Qysy}$ are shown in Fig. 3.7.

When a is very large in comparison to b: $\tanh \alpha_m = \alpha_m$ and $\cosh \alpha_m = \xi = 1$, and consequently, $Q_{ysy}$ becomes diminishingly small which is the case of a composite strip problem.

### 3.4.2 CONTINUOUS COMPOSITE PLATE SUBJECTED TO CONCENTRATED LOADS AT THE CENTER OF ITS PANELS

When a central concentrated load is applied on either panel of the continuous plate (Fig. 3.6), the deflection, shear forces and bending moments can be determined from the following formulas:

$$w^i = \frac{b^{i2} \, p^j}{D} \, K_w^{ij} \tag{3.45}$$

$$Q_x^i = \frac{p^j}{b^i} \, K_{Qx}^{ij} \tag{3.46}$$

$$Q_{yn}^i = \frac{p^j}{b^i} \, K_{Qyn}^{ij} \tag{3.47}$$

$$Q_{yf}^i = \frac{p^j}{b^i} \, K_{Qyf}^{ij} \tag{3.48}$$

$$M_x^i = p^j \, K_{mx}^{ij} \tag{3.49}$$

$$M_y^i = p^j \, K_{my}^{ij} \tag{3.50}$$

$$M_c = p^j \, K_{mc}^j$$

in which

$$j = r \quad \text{means} \quad P^r = 1 \quad \text{and} \quad P^\ell = 0, \quad \text{thus} \quad \phi_{mc}^{j\,'} = \left(\phi_{mc}'\right)_{\substack{P^r=1 \\ P^\ell=0}}$$

$$= \ell \quad \text{means} \quad P^r = 0 \quad \text{and} \quad P^\ell = 0, \quad \text{thus} \quad \phi_{mc}^{j\,'} = \left(\phi_{mc}'\right)_{\substack{P^r=0 \\ P^\ell=1}}$$

$$K_w^{ij} = \sum_{m=1,3,\ldots}^{\infty} \frac{-2}{\pi^3} \frac{(-1)^{(m+1)/2}}{\alpha_m^i \, \xi_m^i} \frac{\phi_{mc}^{j\,'}}{\Psi_m'} R^{r2} R^\ell \frac{\tanh \alpha_m^i}{\cosh \alpha_m^i}$$

$$K_{Qx}^{ij} = \sum_{m=1,3,\ldots}^{\infty} \frac{32}{\pi^3} \frac{\alpha_m^i}{\xi_m^i} \frac{\phi_{mc}^{j\,'}}{\Psi_m'} k \, R^{r3} \frac{1}{\cosh \alpha_m^i}$$

$$K_{Qyn}^{ij} = \pm \sum_{m=1,3,\ldots}^{\infty} \frac{-32}{\pi^3} (-1)^{(m+1)/2} \frac{\alpha_m^i}{\xi_m^i} \frac{\phi_{mc}^{j\,'}}{\Psi_m'} k \, R^{r3}$$

$$(\tanh \alpha_m^i + \coth \alpha_m^i)$$

$$K_{Qyf}^{ij} = \pm \sum_{m=1,3,\ldots}^{\infty} \frac{-32}{\pi^3} (-1)^{(m+1)/2} \frac{\alpha_m^i}{\xi_m^i} \frac{\phi_{mc}^{j\,'}}{\Psi_m'} k \, R^{r3}$$

$$(-\tanh \alpha_m^i + \coth \alpha_m^i)$$

$$K_{mx}^{ij} = \sum_{m=1,3,\ldots}^{\infty} \frac{-8}{\pi^3} (-1)^{(m+1)/2} \frac{R^{\ell 3}}{\xi_m^i \, k^2} \frac{\phi_{mc}^{j\,'}}{\Psi_m'} \frac{1}{\cosh \alpha_m^i} \Big[$$

$$(1-\nu) \, \alpha_m^i \, \tanh \alpha_m^i + 2\nu + 8 \, \alpha_m^{i2} \, \phi^i \, (\nu-1)\Big]$$

$$K_{my}^{ij} = \sum_{m=1,3,\ldots}^{\infty} \frac{-8}{3} (-1)^{(m+1)/2} \frac{R^{\ell 3}}{\xi_m^i k^2} \frac{\phi_{mc}^{j\,'}}{\psi_m'} \frac{1}{\cosh \alpha_m^i} [$$

$$(\nu-1) \alpha_m^i \tanh \alpha_m^i + 2 + 8 \alpha_m^{i2} \phi^i (1-\nu)]$$

$$K_{mc}^{j} = \sum_{m=1,3,\ldots}^{\infty} \frac{-32}{\pi^3} (-1)^{(m+1)/2} k R^{r3} \frac{\phi_{mc}^{j\,'}}{\psi_m'}$$

where the positive sign of ± is for the right panel while the negative sign is for the left panel. Numerical values for these factors are tabulated in Tables B.2 to C.43.

### 3.4.3　CONTINUOUS COMPOSITE PLATE SUBJECTED TO UNIFORMLY DISTRIBUTED LOADS

The practical formulas in this case are given by:

$$w^i = \frac{b^{i4}}{D} p_o^j K_w^{ij} \tag{3.51}$$

$$Q_x^i = b^i p_o^j K_{Qx}^{ij} \tag{3.52}$$

$$Q_{yn}^i = b^i p_o^j K_{Qyn}^{ij} \tag{3.53}$$

$$Q_{yf}^i = b^i p_o^j K_{Qyf}^{ij} \tag{3.54}$$

$$M_x^i = b^{i2} p_o^j K_{mx}^{ij} \tag{3.55}$$

$$M_y^i = b^{i2} p_o^j K_{my}^{ij} \qquad (3.56)$$

$$M_c = b^{r2} p_o^j K_{mc}^j$$

in which

$$j = r \quad \text{means} \quad p_o^r = 1 \quad \text{and} \quad p_o^\ell = 0$$
$$\ell \quad \text{means} \quad p_o^r = 0 \quad \text{and} \quad p_o^\ell = 1$$

$$K_w^{rj} = \sum_{m=1,3,\ldots}^{\infty} \frac{-8}{\pi^5} (-1)^{(m+1)/2} \frac{1}{\alpha_m^r \xi_m^r} \frac{\theta_{mu}^{j\,'}}{\psi_m'} \, k \, R^{r4} \frac{\tanh \alpha_m^r}{\cosh \alpha_m^r}$$

$$K_w^{\ell j} = \sum_{m=1,3,\ldots}^{\infty} \frac{-8}{\pi^5} (-1)^{(m+1)/2} \frac{1}{\alpha_m^\ell \xi_m^\ell} \frac{\phi_{mu}^{j\,'}}{\psi_m'} \frac{R^{\ell 4}}{k} \frac{\tanh \alpha_m^\ell}{\cosh \alpha_m^\ell}$$

$$K_{Qx}^{rj} = \sum_{m=1,3,\ldots}^{\infty} \frac{128}{\pi^5} \frac{\alpha_m^r}{\xi_m^r} \frac{\phi_{mu}^{j\,'}}{\psi_m'} \, k \, R^{r4} \frac{1}{\cosh \alpha_m^r}$$

$$K_{Qx}^{\ell j} = \sum_{m=1,3,\ldots}^{\infty} \frac{128}{\pi^5} \frac{\alpha_m^\ell}{\xi_m^\ell} \frac{\phi_{mu}^{j\,'}}{\psi_m'} \frac{R^{\ell 4}}{k} \frac{1}{\cosh \alpha_m^\ell}$$

$$K_{Qyn}^{rj} = \sum_{m=1,3,\ldots}^{\infty} \frac{-128}{\pi^5} (-1)^{(m+1)/2} \frac{\alpha_m^r}{\xi_m^r} \frac{\phi_{mu}^{j\,'}}{\psi_m'} \, k \, R^{r4} \, ($$

$$\tanh \alpha_m^r + \coth \alpha_m^r)$$

$$K_{Qyn}^{\ell j} = \sum_{m=1,3,\ldots}^{\infty} \frac{128}{\pi^5} (-1)^{(m+1)/2} \frac{\alpha_m^{\ell}}{\xi_m^{\ell}} \frac{\phi_{mu}^{j\,'}}{\psi_m^{'}} \frac{R^{\ell 4}}{k} ($$

$$\tanh \alpha_m^{\ell} + \coth \alpha_m^{\ell})$$

$$K_{Qyf}^{rj} = \sum_{m=1,3,\ldots}^{\infty} \frac{-128}{\pi^5} (-1)^{(m+1)/2} \frac{\alpha_m^{r}}{\xi_m^{r}} \frac{\phi_{mu}^{j\,'}}{\psi_m^{'}} k\, R^{r4} ($$

$$\coth \alpha_m^{r} - \tanh \alpha_m^{r})$$

$$K_{Qyf}^{\ell j} = \sum_{m=1,3,\ldots}^{\infty} \frac{-128}{\pi^5} (-1)^{(m+1)/2} \frac{\alpha_m^{\ell}}{\xi_m^{\ell}} \frac{\phi_{mu}^{j\,'}}{\psi_m^{'}} \frac{R^{\ell 4}}{k} ($$

$$\tanh \alpha_m^{\ell} - \coth \alpha_m^{\ell})$$

$$K_{mx}^{rj} = \sum_{m=1,3,\ldots}^{\infty} \frac{-32}{\pi^5} (-1)^{(m+1)/2} \frac{k\, R^{r4}}{\xi_m^{r}} \frac{\phi_{mu}^{j\,'}}{\psi_m^{'}} \frac{1}{\cosh \alpha_m^{r}} [$$

$$(1-\nu)\; \alpha_m^{r}\; \tanh \alpha_m^{r} + 2\,\nu + 8\,\alpha_m^{r2}\, \phi^r\, (\nu-1)]$$

$$K_{mx}^{\ell j} = \sum_{m=1,3,\ldots}^{\infty} \frac{-32}{\pi^5} (-1)^{(m+1)/2} \frac{R^{\ell 4}}{k\,\xi_m^{\ell}} \frac{\phi_{mu}^{j\,'}}{\psi_m^{'}} \frac{1}{\cosh \alpha_m^{\ell}} [$$

$$(1-\nu)\; \alpha_m^{\ell}\; \tanh \alpha_m^{\ell} + 2\,\nu + 8\,\alpha_m^{\ell 2}\, \phi^{\ell}\, (\nu-1)]$$

$$K_{my}^{rj} = \sum_{m=1,3,\ldots}^{\infty} \frac{-32}{\pi^5} (-1)^{(m+1)/2} \frac{k\ R^{r4}}{\xi_m^r} \frac{\phi_{mu}^{j\,'}}{\psi_m^{'}} \frac{1}{\cosh \alpha_m^r} [$$

$$(\nu-1)\ \alpha_m^r\ \tanh \alpha_m^r + 2 + 8\ \alpha_m^{r2}\ \phi^r\ (1-\nu)]$$

$$K_{my}^{\ell j} = \sum_{m=1,3,\ldots}^{\infty} \frac{-32}{\pi^5} (-1)^{(m+1)/2} \frac{R^{\ell 4}}{k\ \xi_m^\ell} \frac{\phi_{mu}^{j\,'}}{\psi_m^{'}} \frac{1}{\cosh \alpha_m^\ell} [$$

$$(\nu-1)\ \alpha_m^\ell\ \tanh \alpha_m^\ell + 2 + 8\ \alpha_m^{\ell 2}\ \phi^\ell\ (1-\nu)]$$

$$K_{mc}^{j} = \sum_{m=1,3,\ldots}^{\infty} \frac{-128}{\pi^5} (-1)^{(m+1)/2} \frac{R^{\ell 2}\ R^{r2}}{k} \frac{\phi_{mu}^{j\,'}}{\psi_m^{'}}$$

Numerical values for these factors are tabulated in Tables B.44 to B.85.

## 3.5 CONTINUOUS COMPOSITE PLATES BY FINITE DIFFERENCE METHOD

In the previous chapter, the finite difference technique was effectively applied in conjunction with the partial deflection theory to solve simply supported composite plates. The same approach may be used to determine a numerical solution for the deflection of continuous composite plates subjected to lateral loads.

For this purpose, consider for example the continuous plate in Fig. 3.8(a) which consists of two square panels of equal spans, and subjected to a uniformly distributed load on the left one. The finite difference mesh used is shown in Fig. 3.8(b).

By applying the funicular polygon's stencil (Fig. 2.5) on the partial deflection equation (2.73), the bending deflection ordinates at the finite difference mesh points are determined from Equ. (2.76) as:

$$\begin{bmatrix} w_{b1} \\ \vdots \end{bmatrix} \qquad \begin{bmatrix} 0.488 \\ \vdots \end{bmatrix}$$

$$
\begin{Bmatrix}
w_{b2} \\
w_{b3} \\
w_{b4} \\
w_{b5} \\
w_{b6} \\
-w_{b7} \\
-w_{b8} \\
-w_{b9} \\
-w_{b10}
\end{Bmatrix}
= p_o\text{''}\lambda^4/D
\begin{Bmatrix}
0.677 \\
0.488 \\
0.677 \\
0.941 \\
0.677 \\
0.407 \\
0.208 \\
0.573 \\
0.294
\end{Bmatrix}
\tag{3.57}
$$

in which

$\lambda$ =   the finite difference mesh width

=   $a/4$.

where the coefficient and load matrices in Equ. (2.76) are given by:

|   | $w_{b1}$ | $w_{b2}$ | $w_{b3}$ | $w_{b4}$ | $w_{b5}$ | $w_{b6}$ | $w_{b7}$ | $w_{b8}$ | $w_{b9}$ | $w_{b10}$ |
|---|---|---|---|---|---|---|---|---|---|---|
| 1 | 18. | -576. | 72. | -608. | -32. | 32. | 0. | 0. | 0. | 0. |
| 2 | -576. | 1872. | -576. | -32. | -576. | -32. | 0. | 0. | 0. | 0. |
| 3 | 72. | -576. | 1800. | 32. | -32. | -608. | 0. | 0. | 0. | 0. |
| 4 | -1216. | -64. | 64. | 1800. | -576. | 72. | 0. | 0. | 0. | 0. |
| 5 | -64. | -1152. | -64. | -576. | 1872. | -576. | 0. | 0. | 0. | 0. |
| 6 | 64. | -64. | -1216. | 72. | -576. | 1800. | 0. | 0. | 0. | 0. |
| 7 | 0. | 0. | -576. | 0. | 0. | -32. | 1872. | -576. | -576. | -32. |
| 8 | 0. | 0. | 72. | 0. | 0. | 32. | -576. | 1800. | -32. | -608. |
| 9 | 0. | 0. | -64. | 0. | 0. | -576. | -1152. | -64. | 1872. | -576. |
| 10 | 0. | 0. | 64. | 0. | 0. | 72. | -64. | -1216. | -576. | 1800. |

$[A]$ =

$$
\{P_b\} = \lambda^4 p_o\text{''}/D
\begin{Bmatrix}
103.3 \\
120.3 \\
103.3 \\
120.2 \\
138.6 \\
120.2 \\
0. \\
0. \\
0. \\
0.
\end{Bmatrix}
$$

in which

$$\lambda \quad = \quad \text{the finite difference mesh width (Fig. 3.8(b))}$$
$$= \quad a/4.$$

In a similar manner, by applying the higher approximation stencil (Fig. 2.4) on the partial deflection equation (2.74), the shear deflection ordinates at the finite difference mesh points are determined from Equ. (2.78) as:

$$\begin{Bmatrix} w_{s1} \\ w_{s2} \\ w_{s3} \\ w_{s4} \\ w_{s5} \\ w_{s6} \end{Bmatrix} = 12\lambda^2 p_{o"}/S \begin{Bmatrix} 0.0538 \\ 0.0696 \\ 0.0538 \\ 0.0696 \\ 0.0909 \\ 0.0696 \end{Bmatrix} \tag{3.58}$$

where the coefficient and load matrices in Equ. (2.78) are given by:

$$[B] = \begin{bmatrix} w_{s1} & w_{s2} & w_{s3} & w_{s4} & w_{s5} & w_{s6} & \\ -59. & 16. & -1. & 16. & 0. & 0. & 1 \\ 16. & -60. & 16. & 0. & 16. & 0. & 2 \\ -1. & 16. & -59. & 0. & 0. & 16. & 3 \\ 32. & 0. & 0. & -59. & 16. & -1. & 4 \\ 0. & 32. & 0. & 16. & -60. & 16. & 5 \\ 0. & 0. & 32. & -1. & 16. & -59. & 6 \end{bmatrix}$$

$$\{P_s\} = -12\lambda^2 p_{o"}/S \begin{Bmatrix} 1. \\ 1. \\ 1. \\ 1. \\ 1. \\ 1. \end{Bmatrix}$$

The total deflection at point 5 of the finite difference mesh (Fig. 2.8(b)) can be determined by combining the bending and shear components. For a shear parameter equal to 0.1, it is found from Equs. (3.57) and (3.58) that:

$$w_5 = (104.9 \times 10^{-4})(p_{o"}a^4/D) \tag{3.59}$$

The analytical solution for the central deflection of the left panel of the continuous plate in Fig. 3.8(a) is determined by applying the principle of superposition:

$$w_5 = w_{5ss} + w_{5cm}$$

in which

$w_{5ss}$ = the central deflection of a simply supported composite plate having the same properties as the left panel of the continuous plate in Fig. 3.8 (a), and subjected to the same load.

$w_{5cm}$ = the central deflection of the left panel of the continuous composite plate in Fig. 3.8(a) due to the redundant moment.

Thus, by using the practical formulas in Equs. (2.37) and (3.51), an expression for $w_5$ is obtained as

$$w_5 = (K_{wb} + \phi^\ell K_{ws} + K_w^{\ell\ell}) \frac{p_o^\ell a^4}{D}$$

With the numerical values for $K_{wb}$ and $K_{ws}$ from Table A.1, and for $K_w$ "" from Table B.81, noting that $\phi$" = 0.1, it is found that:

$$w_5 = (112.7 \times 10^{-4})(p_o"a^4/D) \tag{3.60}$$

Comparison of this result with that in Equ. (3.59) shows a reasonable agreement.

On the other hand, if the continuous plate (Fig. 3.8(a)) is rigid in shear, the deflection $w_5$ is determined from the finite difference results in Equ. (3.57) as:

$$w_5 = (36.8 \times 10^{-4})(p_o"a^4/D)$$

and from the analytic solution in Equs. (2.37) and (3.51) as

$$w_5 = (K_{wb} + K_w"")(p_o"a^4/D)$$

$$= (35.0 \times 10^{-4})(p_o"a^4/D)$$

where $K_w$ "" is obtained from Table B.80. The agreement in this case is also good.

## 3.6   REFERENCES

[1]   Chong, K.P., Wang, K.A., and Griffith, G.R., "Analysis of Continuous Sandwich Panels in Building Systems", Building and Environment, Vol. 44, 1979.

[2]   Fazio, P.P., and Salahuddin, A., "Analysis of Sandwich Beams Continuous over Supports Subject to Settlement", Department of Civil Engineering, Sir George Williams University, Report No. IV, May 1970.

[3]   Jaeger, L.G., Elementary Theory of Elastic Plates, Pergamon Press, 1964.

[4]   Szilard, R., Theory and Analysis of Plates - Classical and Numerical Methods, Prentice-Hall, Inc., 1974.

[5]   Timoshenko, S., and Krieger, S., Thoery of Plates and Shells, Second edition, McGraw-Hill Book Cimpany, 1959.

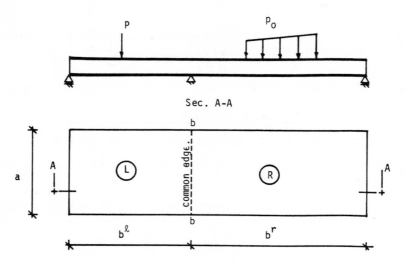

Sec. A-A

(a)  CONTINUOUS COMPOSITE PLATE

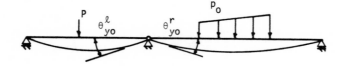

(b)  HINGED PANELS SUBJECTED TO EXTERNAL LOADS

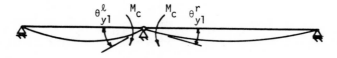

(c)  HINGED PANELS SUBJECTED TO REDUNDANT MOMENT

Fig. 3.1   THE FORCE METHOD OF ANALYSIS

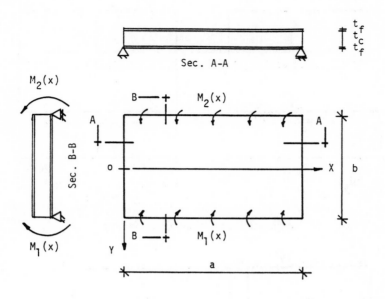

Sec. A-A

$M_2(x)$

$M_2(x)$

$M_1(x)$

$M_1(x)$

Fig. 3.2   RECTANGULAR COMPOSITE PLATE   UNDER MOMENTS
DISTRIBUTED ALONG PARALLEL EDGES

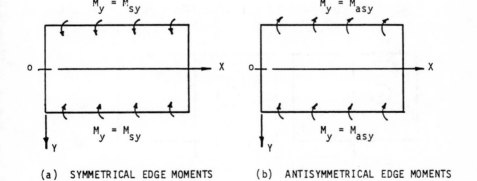

(a)  SYMMETRICAL EDGE MOMENTS          (b)  ANTISYMMETRICAL EDGE MOMENTS

Fig.  3.3   TWO CASES OF MOMENTS DISTRIBUTED ALONG COMPOSITE
PLATE EDGES

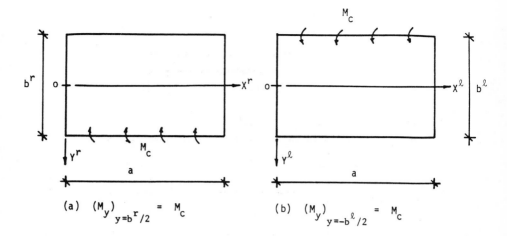

(a) $(M_y)_{y=b^r/2} = M_c$      (b) $(M_y)_{y=-b^\ell/2} = M_c$

Fig. 3.4   COMPOSITE PLATE UNDER EDGE MOMENTS

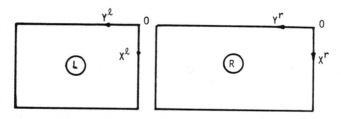

(a)   SIMPLY SUPPORTED PLATE UNDER LATERAL LOAD

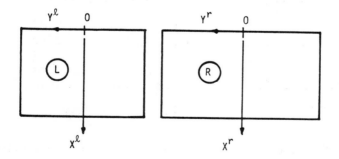

(b)   COMPOSITE PLATE UNDER EDGE MOMENT

Fig. 3.5    COORDINATE AXES FOR COMPOSITE PANELS

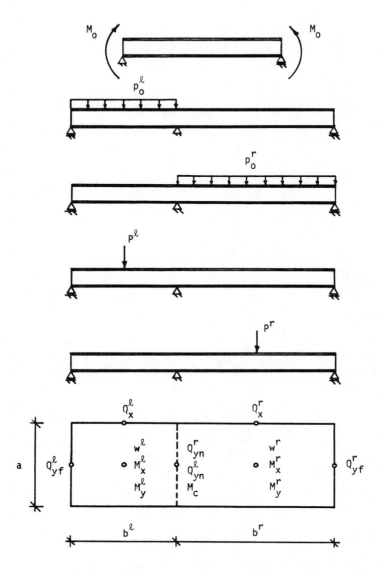

Fig. 3.6 - LOADING TYPES AND THE NOTATIONS USED IN
THE PRACTICAL FORMULAS

**93**

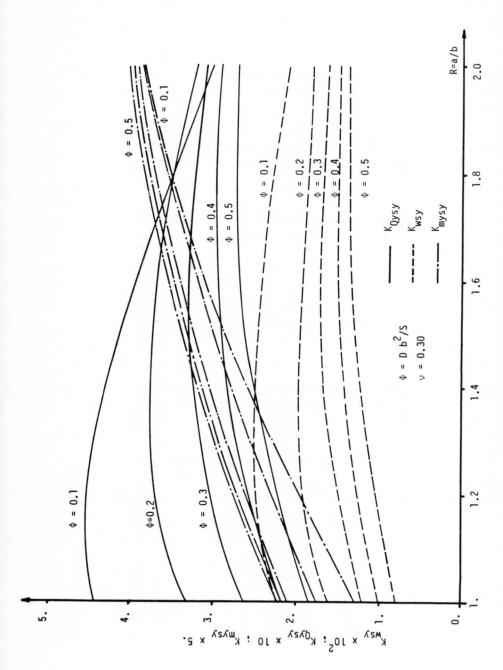

Fig. 3.7  –  NUMERICAL VALUES FOR THE FACTORS IN EQUS. (3.40), (3.42), and (3.44)

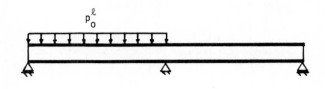

Sec. A-A

(a) LATERAL LOAD

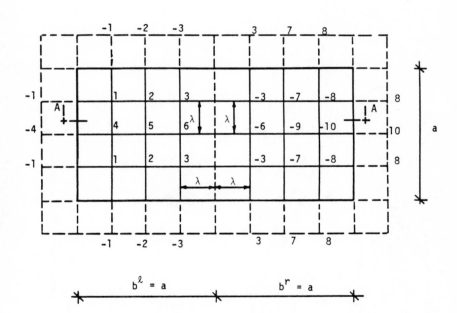

(b) NUMBERING OF FINITE DIFFERENCE MESH POINTS

Fig. 3.8 CONTINUOUS COMPOSITE PLATE BY THE FINITE
DIFFERENCE METHOD

# IV

# Composite Plates and Beam-Columns with Interlayer Elastic Deformations

## 4.1 INTRODUCTION

Composite plates were treated in the preceding chapter with the view to determine their responses to different types of loadings and boundary conditions. The basically three-dimensional problem was reduced to a two-dimensional one by introducing a number of assumptions concerning stress or deformation distribution. One of the basic assumptions was that the constituent material layers are assembled with bondings whose stiffnesses are such as to prevent any interlayer deformations. This ideal behavior of bending materials may not be achieved in practice, since bondings have a certain degree of flexibility. As the response of a composite construction depends on the stiffnesses of its constituents, the finite stiffness of the bonding material should thus be included in the analysis.

The problem of interlayer deformation in other structural systems has been investigated by a number of authors. With regard to layered wood systems, there are numerous theories well established and available in the literature. Amana and Booth [2], Goodman et al. [6, 7, 8, 23], Kuenzi and Wilkinson [13] conducted analytical and experimental studies on layered beams of various materials. In the first study, three types of plywood components were considered: T-beam type; double skin, double rib type; and double skin, multiple rib type of construction. In the second study, two types of wood components were considered: layered beams of three layers, and wood joist floor systems. In the third study, composite beams were considered. In all these studies, the material layers were assembled with nails or by gluing their ends, and although the interlayer slip in these systems was accounted for in the theoretical model, transverse shear deformations were neglected. The finite element method was used by Thompson et al. [20, 21] to study layered wood systems with interlayer slip. In this study, the total strain energy of a layered beam was obtained from the contribution of bending and axial deformations in each layer and slip deformation at the interfaces. On the other hand, interlaminar shear in composites under plane stress was investigated analytically by Puppo and Evensen [19], and with the finite element method by Isakson and Levy [12].

In this chapter, the effect of interlayer elastic deformations on the responses of

97

composite plates and beam columns is investigated analytically. The theories presented next are original. The objective is to provide a means for a more rational and efficient design in terms of structural performance.

## 4.2 SIMPLY SUPPORTED COMPOSITE PLATES UNDER LATERAL LOADS

Consider a plate composed of two faces each of thickness $t_f$ and a core of thickness $t_c$ as shown in Fig. 4.1. The stress state in the faces and core elements is shown in Fig. 4.2.

As the need to consider interlayer deformations necessitates the consideration of compatibility of strains and stresses at the interfaces, the problem is irreducibly three-dimensional and therefore is more complex than before. The assumptions mentioned in Chapter II can still be retained, and new ones are introduced. These are:

(1) The faces are thin in comparison with the core depth. This implies that the flexural rigidity of each face about its own middle surface is negligible, and consequently, the inplane stresses resisted by each skin are uniformly distributed across its thickness. However, the facing material is still much stiffer than the core material, thus any inplane applied loading is resisted mainly by the faces.

(2) The two facings are of equal thickness and made of the same material.

(3) Materials are linearly elastic.

(4) Deformations are small. The deformations in the direction of layer thicknesses are negligible.

(5) Poisson's ratio of the core material, and the inplane shear stress $\tau_{cxy}$ in the core can be ignored.

(6) Interlayer deformations are proportional to the interlayer shears. Thus:

$$\Delta_x = q_x/K_x \qquad\qquad (4.1)$$
$$\Delta_y = q_y/K_y$$

in which

$\Delta_i$ = interlayer deformation in i- direction

$K_i$ = stiffness of the employed bonding in the i- direction. This value can be determined experimentally from the shear test described by Kuenzi and Wilkinson [13].

$q_i$ = interlayer shear stress in the i- direction.

i = subscript denoting x or y.

It is noted that the last two assumptions are new. Assumption (5) effectively implies the existence of a simplified three dimensional stress state in the core. The non-zero stress components are $\sigma_{cx}$, $\sigma_{cy}$, $\tau_{cxz}$, $\tau_{cyz}$ and each need not be uniform across the core thickness. Thus, the associated displacement components $u_c$ along x and $v_c$ along y become functions of x, y, and z. The effect of ignoring the Poisson's ratio is not at all serious in the case of light weight core materials [4, 5, 9, 10, 14, 15, 26]. Assumption (6) represents an idealized form of behavior of the bonding layer. This form has been used by other investigators whose works have been discussed in the introduction of this chapter.

### 4.2.1 GENERAL

When a plate is subjected to applied loads, the different layers tend to slide over one another, and thus interlayer shear stresses are developed. The magnitude of these shear stresses at the interfaces is such as to maintain the compatibility of stresses and deformations.

Now consider the facings, which are assumed to be in a plane stress condition with the non-zero stress components $\sigma_{fx}$, $\sigma_{fy}$, and $\tau_{xy}$. The standard differential equations of equilibrium for the facings (Fig. 4.2) are

$$\frac{\partial \sigma_{fx}^j}{\partial x} + \frac{\partial \tau_{yx}^j}{\partial y} \mp \frac{q_x}{t_f^j} = 0 \tag{4.2}$$

$$\frac{\partial \sigma_{fy}^j}{\partial y} + \frac{\partial \tau_{xy}^j}{\partial x} \mp \frac{q_x}{t_f^j} = 0 \tag{4.3}$$

in which
  j = superscript denoting stresses in the face j, while $t_f^j$ is the thickness of this face;
  f = subscript denoting face.
The negative and positive signs correspond to j = 1 for the bottom face, and j = 2 for the upper one.

The state of stresses in the core must satisfy the following equilibrium equations

$$\frac{\partial \sigma_{cx}}{\partial x} + \frac{\partial \tau_{czx}}{\partial z} = 0$$

$$\tag{4.4}$$

$$\frac{\partial \sigma_{cy}}{\partial y} + \frac{\partial \tau_{czy}}{\partial z} = 0$$

in which

$\sigma_{ci}$ = normal stress in the core in the i- direction;

$\tau_{czi}$ = shear stress in the core.

or in terms of the displacement components

$$E_{cx} \frac{\partial^2 u_c}{\partial x^2} + G_{cx} \frac{\partial^2 u_c}{\partial z^2} = 0 \qquad (4.5)$$

$$E_{cy} \frac{\partial^2 v_c}{\partial y^2} + G_{cy} \frac{\partial^2 v_c}{\partial z^2} = 0 \qquad (4.6)$$

in which

$u_c$, $v_c$ = displacement in the core along the x- and y- directions respectively
E, G  = the elastic and shear modulii of the core material
c   = subscript denoting core.

The stress components in the facings and core must also satisfy the equilibrium equation in terms of the applied load, which is

$$\frac{\partial^2 M_x}{\partial x^2} - 2 \frac{\partial^2 M_{xy}}{\partial x \, \partial y} + \frac{\partial^2 M_y}{\partial y^2} = -p \qquad (4.7)$$

in which

$M_x$, $M_y$, and $M_{xy}$ = the applied moments;
p         = the applied load intensity.

At the interfaces between the core and the faces, the stresses and strains must be compatible. The compatibility equations in terms of stresses are:

$$q_x = G_{cx} \left( \frac{\partial u_c}{\partial z} + \frac{\partial w}{\partial x} \right)_{z=t_c/2}$$

$$q_y = G_{cy} \left( \frac{\partial v_c}{\partial z} + \frac{\partial w}{\partial y} \right)_{z=t_c/2}$$

(4.8)

in which
  $t_c$ = the core depth;
  $w$ = the lateral deflection.

In terms of strains, the compatibility equations are written as [2, 6, 7, 23]:

$$\frac{\partial \Delta_x}{\partial x} = \varepsilon_{fx} - \left( \varepsilon_{cx} \right)_{z=t_c/2}$$

(4.9)

$$\frac{\partial \Delta_y}{\partial x} = \varepsilon_{fy} - \left( \varepsilon_{cy} \right)_{z=t_c/2}$$

(4.10)

in which
  $\varepsilon_{fi}$ = normal strain in the face along the i- direction
    = $\partial u_f/\partial x$ along x- direction, and $\partial v_f/\partial y$ along y- direction;
  $\varepsilon_{ci}$ = normal strain in the core along the i- direction
    = $\partial u_c/\partial x$ along x- direction, and $\partial v_c/\partial y$ along y- direction.

Solutions to the problem must also satisfy prescribed boundary conditions. With respect to the simply supported plates subjected to transverse loads (Fig. 4.1), the relevant boundary conditions are:
  (i) No deflection or normal stresses should exist at the plate edges, thus
    at x = 0, 2a    $\sigma_{fx} = 0$ and $\sigma_{cx} = \gamma_y = 0$ and w = 0    (4.11)
    at y = 0, 2b    $\sigma_{fy} = 0$ and $\sigma_{cy} = \gamma_x = 0$ and w = 0
    in which
    2a, 2b  =  the plate dimensions in x- and y- directions, respectively;
    $\gamma i$  =  the shear strain in the plane zi.
  (ii) As the two facings are identical, the middle plane of the plate remains unstrained under loads, thus:

at z = 0    $u_c = v_c = 0$    (4.12)

(iii) For symmetrical loading about the plate center lines, the inplane displacement components vanish along those lines, thus:

at x = a    $u_f = u_c = 0$    (4.13)

at y = b    $v_f = v_c = 0$

in which

$u_f, v_f$ = the displacement in the facings along x- and y- directions, respectively.

## 4.2.2    ANALYTIC SOLUTIONS FOR SIMPLY SUPPORTED COMPOSITE PLATES

Consider the simply supported plate in Fig. 4.1, made of two faces each of thickness $t_f$ and a core of thickness $t_c$. The displacement components $u_f$ and $v_f$ which satisfy u = 0 and v = 0 in Equ. (4.13) are taken as [3, 16, 17, 24, 25]:

$$u_f = \sum_{m=1,3,\ldots}^{\infty} \sum_{n=1,3,\ldots}^{\infty} A_{mn} \cos \alpha_m x \sin \beta_n y \qquad (4.14)$$

$$v_f = \sum_{m=1,3,\ldots}^{\infty} \sum_{n=1,3,\ldots}^{\infty} B_{mn} \sin \alpha_m x \cos \beta_n y \qquad (4.15)$$

in which

$\alpha_m = m\pi/2a$

$\beta_n = n\pi/2b$

$A_{mn}, B_{mn}$ = unknown coefficients.

From Equs. (4.14) and (4.15), expressions for the stresses in the facings can be written as:

$$\sigma_{fx} = \frac{E_f}{\nu^2 - 1} \sum_{m=1,3,\ldots}^{\infty} \sum_{n=1,3,\ldots}^{\infty} (A_{mn} \alpha_m + \nu B_{mn} \beta_n)$$

$$\sin \alpha_m x \sin \beta_n y \qquad (4.16)$$

$$\sigma_{fy} = -\frac{E_f}{\nu^2 - 1} \sum_{m=1,3,\ldots}^{\infty} \sum_{n=1,3,\ldots}^{\infty} (B_{mn}\beta_n + \nu A_{mn}\alpha_m)$$
$$\sin\alpha_m x \sin\beta_n y \qquad (4.17)$$

$$\tau_{xy} = \frac{E_f}{2(1+\nu)} \sum_{m=1,3,\ldots}^{\infty} \sum_{n=1,3,\ldots}^{\infty} (A_{mn}\beta_n + B_{mn}\alpha_m)$$
$$\cos\alpha_m x \cos\beta_n y \qquad (4.18)$$

in which

$E_f$ = elastic modulus of the face material

$\upsilon$ = Poisson's ratio of the facings material

It is noted that the expressions in Equs. (4.16), (4.17), and (4.18) satisfy the boundary conditions $\sigma_{fx} = 0$ and $\sigma_{fy} = 0$ in Equ. (4.11).

Expressions for the interlayer shear stresses $q_x$ and $q_y$ are obtained by differentiating the above stresses in accordance with the equilibrium equations (4.2) and (4.3) as:

$$q_x = \frac{E_f t_f}{1+\nu} \sum_{m=1,3,\ldots}^{\infty} \sum_{n=1,3,\ldots}^{\infty} \left( \frac{A_{mn}\alpha_m^2 + \nu B_{mn}\beta_n\alpha_m}{\nu - 1} \right.$$
$$\left. - \frac{A_{mn}\beta_n^2 + B_{mn}\alpha_m\beta_n}{2} \right) \cos\alpha_m x \sin\beta_n y \qquad (4.19)$$

$$q_y = \frac{E_f t_f}{1+\nu} \sum_{m=1,3,\ldots}^{\infty} \sum_{n=1,3,\ldots}^{\infty} \left( \frac{B_{mn}\beta_n^2 + \nu A_{mn}\alpha_m\beta_n}{\nu - 1} \right.$$
$$\left. - \frac{A_{mn}\beta_n\alpha_m + B_{mn}\alpha_m^2}{2} \right) \sin\alpha_m x \cos\beta_n y \qquad (4.20)$$

On the other hand, a solution for the core displacement satisfying the equilbrium equations (4.5) and (4.6) as well as the boundary conditions $u_c = 0$ and $v_c = 0$ in Equs. (4.12) and (4.13), $\sigma_{cx} = 0$ and $\sigma_{cy} = 0$ in Equs. (4.11) is found as [11]:

$$u_c = \sum_{m=1,3,\ldots}^{\infty} C_m \cos \alpha_m x \sinh \mu_x \alpha_m z \tag{4.21}$$

$$v_c = \sum_{n=1,3,\ldots}^{\infty} D_n \cos \beta_n y \sinh \mu_y \beta_n z \tag{4.22}$$

in which

$$\mu_x = (\frac{E_{cx}}{G_{cx}})^{1/2}$$

$$\mu_y = (\frac{E_{cy}}{G_{cy}})^{1/2}$$

$$C_m, D_n = \text{unknown functions of } y \text{ and } x, \text{ respectively.}$$

The functions $C_m$ and $D_n$ can be expressed in terms of $A_{mn}$ and $B_{mn}$ by using the compatibility equations (4.9) and (4.10). By substituting Equs. (4.1), (4.14), (4.19), and (4.21) in Equ. (4.9) an expression for $C_m$ is obtained as:

$$C_m = \sum_{n=1,3,\ldots}^{\infty} \frac{1}{\sinh(\mu_x \alpha_m t_c/2)} [\frac{E_f t_f}{K_x (1+\nu)} (\frac{A_{mn} \alpha_m^2 + \nu B_{mn} \beta_n \alpha_m}{1 - \nu}$$

$$+ \frac{A_{mn} \beta_n^2 + B_{mn} \alpha_m \beta_n}{2}) + A_{mn}] \sin \beta_n y \tag{4.23}$$

In a similar manner, from Equs. (4.1), (4.15), (4.20), (4.22), and (4.10) it follows that:

$$D_n = \sum_{m=1,3,\ldots}^{\infty} \frac{1}{\sinh(\mu_y \beta_n t_c/2)} [\frac{E_f t_f}{K_y (1+\nu)} (\frac{B_{mn} \beta_n^2 + \nu A_{mn} \alpha_m \beta_n}{1 - \nu}$$

$$+ \frac{A_{mn} \beta_n \alpha_m + B_{mn} \alpha_m^2}{2}) + B_{mn}] \sin \alpha_m x \tag{4.24}$$

The coefficient $B_{mn}$ can be also expressed in terms of $A_{mn}$ by using the compatibility equations of stresses (4.8). The deflection function for simply supported plates which satisfies $w = 0$ in Equ. (4.11) is taken as:

$$w = \sum_{m=1,3,\ldots}^{\infty} W_m \sin \alpha_m x \sin \beta_n y$$

By proper substitution of $w$, $q_x$, and $u_c$ in the first equation of (4.8), and by equating the coefficients of $\cos \alpha_m x \sin \beta_n y$ in both of its sides, it is found that:

$$W_m = \frac{E_f t_f}{G_{cx}(1+\nu)} \left( \frac{A_{mn}\alpha_m + \nu B_{mn}\beta_n}{\nu - 1} - \frac{A_{mn}\beta_n^2 + B_{mn}\alpha_m\beta_n}{2\alpha_m} \right.$$

$$- \mu_x \left( \coth \frac{1}{2} \mu_x \alpha_m t_c \right) \left[ \frac{E_f t_f}{K_x(1+\nu)} \left( \frac{A_{mn}\alpha_m^2 + \nu B_{mn}\beta_n\alpha_m}{1 - \nu} \right. \right.$$

$$\left. \left. \left. + \frac{A_{mn}\beta_n^2 + B_{mn}\alpha_m\beta_n}{2} \right) + A_{mn} \right] \right] \qquad (4.25a)$$

In a similar manner, from Equs. (4.20), (4.22), and the second equation in (4.8), it is found that:

$$W_m = \frac{E_f t_f}{G_{cy}(1+\nu)} \left( \frac{B_{mn}\beta_n + \nu A_{mn}\alpha_m}{\nu - 1} - \frac{A_{mn}\alpha_m\beta_n + B_{mn}\alpha_m^2}{2\beta_n} \right)$$

$$- \mu_y \left( \coth \frac{1}{2} \mu_y \beta_n t_c \right) \left[ \frac{E_f t_f}{K_y(1+\nu)} \left( \frac{B_{mn}\beta_n^2 + \nu B_{mn}\alpha_m\beta_n}{1 - \nu} \right. \right.$$

$$\left. \left. \left. + \frac{A_{mn}\beta_n\alpha_m + B_{mn}\alpha_m^2}{2} \right) + B_{mn} \right] \right] \qquad (4.25b)$$

By equating the right hand side of Equs. (4.25a) and (4.25b), it follows that:

$$B_{mn} = (\phi_{mn}/\psi_{mn}) \, A_{mn} \tag{4.26}$$

in which

$$
\begin{aligned}
\phi_{mn} \;=\; & \frac{E_f \, t_f}{G_{cy} \, (1+\nu)} \; \Big[ -\frac{1}{2} \, \alpha_m + \frac{G_{cy}}{2 \, G_{cx}} \, \frac{\beta_n^2}{\alpha_m} \\[2mm]
& + \frac{G_{cy} \, \mu_y}{2 \, K_y} \, \alpha_m \, \beta_n \, \frac{\nu+1}{\nu-1} \, \coth \tfrac{1}{2} \mu_y \, \beta_n \, t_c \\[2mm]
& + \frac{G_{cy} \, \mu_x}{2 \, K_x} \, \Big( \frac{2 \, \alpha_m^2}{1-\nu} + \beta_n^2 \Big) \, \coth \tfrac{1}{2} \mu_x \, \alpha_m \, t_c \\[2mm]
& + \frac{G_{cy}}{G_{cx}} \, \frac{\alpha_m}{1-\nu} + \frac{\nu \, \alpha_m}{\nu-1} \\[2mm]
& + \frac{G_{cy} \, (1+\nu)}{E_f \, t_f} \, \mu_x \, \coth \tfrac{1}{2} \mu_x \, \alpha_m \, t_c \Big]
\end{aligned}
$$

$$
\begin{aligned}
\psi_{mn} \;=\; & \frac{E_f \, t_f}{G_{cx} \, (1+\nu)} \; \Big[ -\frac{1}{2} \beta_n \, \frac{G_{cx}}{2 \, G_{cy}} \, \frac{\alpha_m^2}{\beta} \\[2mm]
& + \frac{G_{cx} \, \mu_x}{2 \, K_x} \, \alpha_m \, \beta_n \, \frac{\nu+1}{\nu-1} \, \coth \tfrac{1}{2} \mu_x \, \alpha_m \, t_c \\[2mm]
& + \frac{G_{cx} \, \mu_y}{2 \, K_y} \, \Big( \frac{2 \, \beta_n^2}{1-\nu} + \alpha_m^2 \Big) \, \coth \tfrac{1}{2} \mu_y \, \beta_n \, t_c
\end{aligned}
$$

$$+ \frac{G_{cx}}{G_{cy}} \frac{\beta_n}{1-\nu} + \frac{\nu \beta_n}{\nu - 1}$$

$$+ \frac{G_{cx} (1+\nu)}{E_f t_f} \mu_y \coth \frac{1}{2} \mu_y \beta_n t_c ]$$

The only unknown coefficient now is $A_{mn}$. This can be determined by using the equilibrium equation (4.7). The moments $M_x$, $M_y$, and $M_{xy}$ can be expressed in terms of the stresses components as:

$$M_x = \sigma_{fx} t_f h + \int_{-t_c/2}^{t_c/2} \sigma_{cx} z \, dz$$

$$M_y = \sigma_{fy} t_f h + \int_{-t_c/2}^{t_c/2} \sigma_{cy} z \, dz$$

$$M_{xy} = - \tau_{xy} t_f h$$

in which

$$h = t_c + t_f$$

$$\sigma_{cx} = E_{cx} \frac{\partial u_c}{\partial x}$$

$$\sigma_{cy} = E_{cy} \frac{\partial v_c}{\partial y}$$

In terms of these stresses, the equilibrium equation (4.7) becomes:

$$t_f \ h \ (\frac{\partial^2 \sigma_{fx}}{\partial x^2} + 2 \frac{\partial^2 \tau_{xy}}{\partial x \ \partial y} + \frac{\partial^2 \sigma_{fy}}{\partial y^2})$$

$$+ E_{cx} \frac{\partial^2}{\partial x^2} \int_{-t_c/2}^{t_c/2} \frac{\partial u_c}{\partial x} \ z \ dz + E_{cy} \frac{\partial^2}{\partial y^2} \int_{-t_c/2}^{t_c/2} \frac{\partial v_c}{\partial y} \ z \ dz$$

$$= - p$$

By expanding the applied load in double trigonometric series, and by substituting Equs. (4.16), (4.17), (4.18), (4.21), (4.22), and by considering Equs. (4.23), (4.24), and (4.26) in the above equation, an expression for $A_{mn}$ is obtained as:

$$A_{mn} = P_{mn}/\xi_{mn} \tag{4.27}$$

in which

$P_{mn}$ = the coefficient of Fourier expansion of the applied load.

$$\xi_{mn} = \frac{E_f \ t_f \ h}{\nu^2 - 1} \ (\alpha_m^3 + \nu \beta_n \ \alpha_m^2 \ \frac{\phi_{mn}}{\psi_{mn}})$$

$$- \frac{E_f \ t_f \ h}{1 + \nu} \ (\alpha_m \ \beta_n^2 + \alpha_m^2 \ \beta_n \ \frac{\phi_{mn}}{\psi_{mn}})$$

$$+ \frac{E_f \ t_f \ h}{\nu^2 - 1} \ (\nu \alpha_m \ \beta_n^2 + \beta_n^3 \ \frac{\phi_{mn}}{\psi_{mn}})$$

$$+ \frac{2 \ E_{cx} \ \mu_m \ C'_{mn}}{\mu_x^2} \ (\frac{1}{2} \mu_x \ \alpha_m \ t_c \ \cosh \frac{1}{2} \mu_x \ \alpha_m \ t_c$$

$$- \sinh \frac{1}{2} \mu_x \ \alpha_m \ t_c) + \frac{2 \ E_{cy} \ \beta_n \ D'_{mn}}{\mu_y^2} \ ($$

$$\frac{1}{2} \mu_y \ \beta_n \ t_c \ \cosh \frac{1}{2} \mu_y \ \beta_n \ t_c - \sinh \frac{1}{2} \mu_y \ \beta_n \ t_c)$$

$$C'_{mn} = \frac{1}{\sinh \frac{1}{2} \mu_x \alpha_m t_c} \left[ \frac{E_f t_f}{K_x (1+\nu)} \left( \frac{\alpha_m^2 + \nu \alpha_m \beta_n \phi_{mn}/\Psi_{mn}}{\nu - 1} \right) \right.$$

$$\left. - \frac{\beta_n^2 + \alpha_m \beta_n \phi_{mn}/\Psi_{mn}}{2} \right) - 1 \right]$$

$$D'_{mn} = \frac{1}{\sinh \frac{1}{2} \mu_y \beta_n t_c} \left[ \frac{E_f t_f}{K_y (1+\nu)} \left( \frac{\nu \alpha_m \beta_n + \beta_n^2 \phi_{mn}/\Psi_{mn}}{\nu - 1} \right) \right.$$

$$\left. - \frac{\alpha_m \beta_n + \alpha_m^2 \phi_{mn}/\Psi_{mn}}{2} \right) - \frac{\phi_{mn}}{\Psi_{mn}} \right]$$

In the particular case of isotropic core and bonding materials:

$$
\begin{aligned}
G_{cx} &= G_{cy} = G_c \\
E_{cx} &= E_{cy} = E_c \\
K_x &= K_y = K
\end{aligned}
$$

hence

$$\mu_x = \mu_y = (E_c/G_c)^{1/2} = \mu$$

and simplified formulas can be obtained as will be shown in subsequent sections.

## 4.3 PARAMETRIC STUDY OF BONDING EFFECTS ON THE BEHAVIOR OF COMPOSITE PLATES

The complexity of the above solution makes it difficult to see the effects of bonding on the plate responses. To illustrate these effects, a simply supported square plate is considered (Fig. 4.3). The plate is made of two aluminum faces, a plastic foam core, and assembled with an isotropic bonding. The facings and core properties are:
(i) for facings

$$
\begin{aligned}
t_f &= .04 \text{ in.} \\
E_f &= 10^7 \text{ psi} \\
\upsilon &= .33
\end{aligned}
$$

(ii) for core

$$t_c = 2.0 \text{ in.}$$
$$E_c = 2 \times 10^4 \text{ psi}$$
$$G_c = 10 \text{ psi}$$

(iii) for the plate

$$2a = 2b = 40 \text{ in.}$$

In this particular case the expressions developed for $\phi_{mn}$, $\psi_{mn}$, and $\xi_{mn}$ become:

$$
\phi_{mn} = \frac{E_f \, t_f}{G_c \, (1+\nu)} \left[ \frac{1}{2} \alpha_m + \frac{1}{2} \frac{\beta_n^2}{\alpha_m} \right.
$$

$$
+ \frac{G_c \, \mu}{2 \, K} \, \alpha_m \, \beta_n \, \frac{\nu+1}{\nu-1} \, \coth \frac{1}{2} \, \mu \, \beta_n \, t_c
$$

$$
+ \frac{G_c \, \mu}{2 \, K} \, \left( \frac{2 \, \alpha_m^2}{1-\nu} + \beta_n^2 \right) \, \coth \frac{1}{2} \, \mu \, \alpha_m \, t_c
$$

$$
\left. + \frac{G_c \, (1+\nu)}{E_f \, t_f} \, \mu \, \coth \frac{1}{2} \, \mu \, \alpha_m \, t_c \right]
$$

$$
\psi_{mn} = \frac{E_f \, t_f}{G_c \, (1+\nu)} \left[ \frac{1}{2} \beta_n + \frac{1}{2} \frac{\alpha_m^2}{\beta_n} \right.
$$

$$
+ \frac{G_c \, \mu}{2 \, K} \, \alpha_m \, \beta_n \, \frac{\nu+1}{\nu-1} \, \coth \frac{1}{2} \, \mu \, \alpha_m \, t_c
$$

$$
+ \frac{G_c \, \mu}{2 \, K} \, \left( \frac{2 \, \beta_n^2}{1-\nu} + \alpha_m^2 \right) \, \coth \frac{1}{2} \, \mu \, \beta_n \, t_c
$$

$$
+ \frac{G_c \, (1+\nu)}{E_f \, t_f} \, \mu \, \coth \frac{1}{2} \, \mu \, \beta_n \, t_c
$$

$$\xi_{mn} = \frac{E_f t_f h}{\nu^2 - 1} (\alpha_m^3 + \nu\beta_n \alpha_m^2 \frac{\phi_{mn}}{\Psi_{mn}})$$

$$- \frac{E_f t_f h}{1 + \nu} (\alpha_m \beta_n^2 + \alpha_m^2 \beta_n \frac{\phi_{mn}}{\Psi_{mn}})$$

$$+ \frac{E_f t_f h}{\nu^2 - 1} (\nu\alpha_m \beta_n^2 + \beta_n^3 \frac{\phi_{mn}}{\Psi_{mn}})$$

$$+ 2 G_c \alpha_m C_m' (\frac{1}{2} \mu \alpha_m t_c \cosh \frac{1}{2} \mu \alpha_m t_c - \sinh \frac{1}{2} \mu \alpha_m t_c)$$

$$+ 2 G_c \beta_n D_n' (\frac{1}{2} \mu \beta_n t_c \cosh \frac{1}{2} \mu \beta_n t_c - \sinh \frac{1}{2} \mu \beta_n t_c)$$

in which

$$C_{mn}' = \frac{1}{\sinh \frac{1}{2} \mu \alpha_m t_c} [\frac{E_f t_f}{K(1+\nu)} (\frac{\alpha_m^2 + \nu\alpha_m \beta_n \phi_{mn}/\Psi_{mn}}{\nu - 1}$$

$$- \frac{\beta_n^2 + \alpha_m \beta_n \phi_{mn}/\Psi_{mn}}{2}) - 1]$$

$$D_{mn}' = \frac{1}{\sinh \frac{1}{2} \mu \beta_n t_c} [\frac{E_f t_f}{K(1+\nu)} (\frac{\nu\alpha_m \beta_n + \beta_n^2 \phi_{mn}/\Psi_{mn}}{\nu - 1}$$

$$- \frac{\alpha_m \beta_n + \alpha_m^2 \phi_{mn}/\Psi_{mn}}{2}) - \frac{\phi_{mn}}{\Psi_{mn}}]$$

Three loading types are considered: uniformly distributed load of intensity $p_o$ (Fig. 4.3(a)), partial load on a central square area (Fig. 4.3(b)), and a central concentrated load P (Fig. 4.3(c)). Because of the symmetry in the plate geometry and the applied loads:

$$\left. \begin{array}{l} \left( \sigma_{fx} \right)_{\substack{x=a \\ y=a}} = \left( \sigma_{fy} \right)_{\substack{x=a \\ y=a}} = \sigma_f \\[3em] \left( \sigma_{cx} \right)_{\substack{x=a \\ y=a \\ z=t_c/2}} = \left( \sigma_{cy} \right)_{\substack{x=a \\ y=a \\ z=t_c/2}} = \sigma_c \\[3em] \left( q_x \right)_{\substack{x=0,2a \\ y=a \\ z=t_c/2}} = \left( q_y \right)_{\substack{x=a \\ y=0,2a \\ z=t_c/2}} = q \\[3em] \left( \tau_{xz} \right)_{\substack{x=0,2a \\ y=a \\ z=0}} = \left( \tau_{yz} \right)_{\substack{x=a \\ y=0,2a \\ z=0}} = \tau_c \\[3em] \left( \tau_{xy} \right)_{\substack{x=0,2a \\ y=0,2b}} \qquad\quad = \tau_{xy} \end{array} \right\} \qquad (4.28)$$

in which

$\sigma_f, \sigma_c$ = the normal stress in the facings and core, respectively, at the plate center;

$q, \tau_c$ = the interlayer shear stress and transverse shear stress in the core, respectively, at the mid-edges of the plate;

$\tau_{xy}$ = the shear stress in the facings at the plate corners.

These stresses, in addition to the central deflection, are calculated in each load case for a chosen range of bonding stiffnesses from $10^3$ to $10^4$ psi/in. The results are shown graphically in Figs. 4.4, 4.5, 4.6, and 4.7. It is seen that the deflection shows greater sensitivity to the variation of the K value when the latter is in the lower range; and beyond a certain level of bonding stiffness, it can be practically considered as rigid. In the case of uniform loading (Fig. 4.4), for example, a change in K value from 3000 to 2000 psi/in. induces a deflection increase fourteen times greater then that when K changes from 10000 to 9000 psi/in. On the other hand, an increase in the adhesive stiffness is accompanied by a decrease in the normal stress of the core. The resulting loss in the resisting moment is compensated by a slight increase in the faces normal stress. In all cases, the interlaminar shear stress is practically independent of the bonding stiffness.  The effects on the normal and shear stress distributions in the core are shown in Fig. 4.8.  It is seen that by increasing the K value the normal stress, $\sigma_c$, is reduced, and the shear stress, $\tau_c$, changes to linear distribution.

The above discussion has yet to bring an important point.  By using the existing theories [1, 18], normal and shear stresses in the facings, and the transverse shear stress in the core of a plate can be determined with a small margin of error.  The bonding stiffness should be included in the analysis whenever the deflection is the quantity of interest.

## 4.4   BONDING   EFFECTS   ON THE   RESPONSE   OF   COMPOSITE BEAM-COLUMNS

### 4.4.1   GENERAL

A simply supported beam of span 2a is subjected to a uniform load of intensity $p_o$, a concentrated load Q at mid span and axial forces P, as shown in Fig. 4.9(a).  Beam imperfection is assumed in the form of a parabolic distribution as:

$$w_o = 2\,\delta\,(x/a)[1 - (x/a)] \tag{4.29}$$

in which
   $w_o$   =   initial imperfection
   $\delta$   =   amplitude of $w_o$ at mid span (Fig. 4.9(a)).

The assumptions mentioned in section 4.2 can still be retained here.  The behavior of the beam in Fig. 4.9(a) is a special case of that for the plate in Fig. 4.1, where the stresses and deformations are considered only in the transverse plane zx.  Therefore, it does not seem necessary to represent the conditions governing the behavior of the beam in a separate section, and the solution procedure is presented next.

### 4.4.2   ANALYTIC SOLUTIONS FOR COMPOSITE BEAM-COLUMN

The equilibrium of a core element requires that:

$$\frac{\partial \sigma_c}{\partial x} + \frac{\partial \tau}{\partial z} = 0 \tag{4.30}$$

in which

$\sigma_c$ = normal stress in the core;

$\tau$ = transverse shear stress in the core;

x, z = coordinate axes as shown in Fig. 4.9(a).

The stresses are related to the core deformations by

$$\sigma_c = E_c \frac{\partial u}{\partial x} \tag{4.31}$$

$$\tau = G \left(\frac{\partial u}{\partial z} + \frac{\partial w}{\partial x}\right) \tag{4.32}$$

in which

u, w = displacement of a point in the core along x- and z- axes, respectively;

$E_c$, G = elastic and shear modulii of the core material, respectively.

Substituting Equs. (4.31) and (4.32) into Equ. (4.30) and by making use of assumption (4) in section 4.2 results in the following equilibrium equation in terms of displacement:

$$E_c \frac{\partial^2 u}{\partial x^2} + G_c \frac{\partial^2 u}{\partial z^2} = 0 \tag{4.33}$$

It is noted that Equs. (4.30) and (4.33) have the same forms as the first equation in Equ. (4.4) and Equ. (4.5), respectively. Thus, a solution which satisfies the following boundary conditions:

at x = a  and  z = 0     u = 0                                                 (4.34)

at x = 0  and  x = 2a   $\partial u/\partial x = 0$                           (4.35)

and Equ. (4.33) would have the form of Equ. (4.21), that is [11]

$$u = \sum_{m=1,3,\ldots}^{\infty} C_m \cos \alpha_m x \, \sinh \mu \, \alpha_m z \tag{4.36}$$

in which

$\alpha_m$  =  $m\pi/2a$;

$\mu$  =  $(E_c/G)^{1/2}$;

$C_m$  =  unknown coefficient.

At the bonding layer, compatibility of deformation and shear stress must be satisfied. The compatibility equation in terms of normal strains is:

$$\frac{d\Delta}{dx} = \varepsilon_f - (\varepsilon_c)_{z=t_c/2} \tag{4.37}$$

in which

$\Delta$  =  interlayer deformation;

$\varepsilon_i$  =  normal strain of the i- layer;

f, c  =  subscripts denoting face and core, respectively.

and in terms of shear stress is:

$$\frac{q}{b} = G \left(\frac{\partial u}{\partial z} + \frac{\partial w}{\partial x}\right)_{z=t_c/2} \tag{4.38}$$

in which

q  =  interlayer shear flux;

b  =  beam width.

Equations (4.37) and (4.38) are similar to the compatibility equations (4.9) and (4.8), respectively.

The equilibrium equation of a face element (Fig. 4.9(c)) requires that:

$$q = dN/dx \tag{4.39}$$

in which

N  =  normal force in the facing.

According to assumption (6) in section 4.2, the shear flux q would introduce an interlayer deformation of amount:

$$\Delta \;=\; q/k \tag{4.40}$$

in which
  $K$  =  bonding stiffness.

Equations (4.39) and (4.40) give:

$$\frac{d\Delta}{dx} \;=\; \frac{1}{K}\,\frac{d^2N}{dx^2} \tag{4.41}$$

Substitution of Equs. (4.36) and (4.41) into Equ. (4.37), noting that $\varepsilon_a = N/E_f A_f$, yields:

$$\frac{d^2N}{dx^2} \;-\; \frac{K\,N}{E_f\,A_f} \;-\; K\sum_{m=1,3,\dots}^{\infty}\;\alpha_m\,C_m\,\sin\,\alpha_m x\;\sinh\tfrac{1}{2}\mu\,\alpha_m\,t_c \;=\; 0 \tag{4.42}$$

in which
  $A_f$  =  the face area;
     =  $t_f b$, where $t_f$ is the face thickness.

The boundary conditions to be satisfied by N are:

  at $x = 0, 2a$   $N = P/2$                        $\qquad(4.43)$
  at $x = a$      $dN/dx = 0$

Thus, the solution for N may be written as:

$$N \;=\; \frac{1}{2}\,P\,(\cosh\theta x - \tanh\theta a\,\sin\theta x) \;-\; \sum_{m=1,3,\dots}^{\infty}\;N_m\,C_m\,\sin\,\alpha_m x \tag{4.44}$$

in which

$$N_m = \frac{\alpha_m \sinh \frac{1}{2} \mu \alpha_m t_c}{\frac{1}{E_f A_f} + \frac{\alpha_m^2}{K}}$$

$$\theta = (\frac{K}{E_f A_f})^{1/2}$$

For the symmetrical loading shown in Fig. 4.9(a), the deflection curve takes the form:

$$w = \sum_{m=1,3,\ldots}^{\infty} W_m \sin \alpha_m x \qquad (4.45)$$

in which
$W_m$ = unknown coefficients.

The primary unknowns now are $C_m$ and $W_m$. The applied bending moment, M at any section of the beam can be expressed as:

$$M = \frac{Q}{2} x + P_0 ax (1 - \frac{x}{2a}) + P (w + w_0) \qquad (4.46)$$

Substituting Eqs. (4.29) and (4.45) and expressing the result in the form of sine series:

$$M = \sum_{m=1,3,\ldots}^{\infty} [(-1)^{(m+3)/2} \frac{Q}{a \alpha_m^2} + \frac{2 P_0}{a \alpha_m^3} + P (\frac{4 \delta}{a^3 \alpha_m^3} + W_m)] \sin \alpha_m x$$
$$\ldots\ldots (4.47)$$

The resisting moment contributed by the core and facings is:

$$M_r = \int_{-t_c/2}^{t_c/2} \sigma_c z b \, dz + Nh$$

$$= \int_{-t_c/2}^{t_c/2} E_c \frac{\partial u}{\partial x} z b \, dz + Nh \qquad (4.48)$$

in which

$$h = t_c + t_f$$

Equations (4.48) in conjunction with Equs. (4.36) and (4.44) give:

$$M_r = \sum_{m=1,3,\dots}^{\infty} \left[ \frac{h \, P \, \alpha_m}{a \, (\alpha_m^2 + \theta^2)} - h \, N_m \, C_m - M_m \, C_m \right] \sin \alpha_m x \quad (4.49)$$

in which

$$M_m = E_c b \left( \frac{t_c}{\mu} \cosh \frac{1}{2} \mu \, \alpha_m \, t_c - \frac{2}{\mu^2 \, \alpha_m} \sinh \frac{1}{2} \mu \, \alpha_m \, t_c \right)$$

Now equate the coefficients of $\sin \alpha_m x$ in Equs. (4.47) and (4.49) yielding the first equation for the determination of $W_m$ and $C_m$:

$$(-1)^{(m+3)/2} \frac{Q}{a \, \alpha_m^2} + \frac{2 \, P_0}{a \, \alpha_m^3} + P \left( \frac{4 \, \delta}{a^3 \, \alpha_m^3} + W_m \right) =$$

$$\frac{h \, P \, \alpha_m}{a \, (\alpha_m^2 + \theta^2)} - h \, N_m \, C_m - M_m \, C_m \quad (4.50)$$

The second equation required is obtained from the stress compatibility relation expressed by Equ. (4.38), with the appropriate substitution for q, N, u, and w.  For the mth term:

$$\frac{P \, \alpha_m}{G \, a \, b \, (\alpha_m^2 + \theta^2)} = C_m \left( \frac{N_m}{b \, G} + \mu \cosh \frac{1}{2} \mu \, \alpha_m \, t_c \right) + W_m \quad (4.51)$$

Solving for $C_m$ from Equs. (4.50) and (4.51) yields:

$$C_m = A_m'/A_m'' \quad (4.52)$$

in which

$$A'_m = (-1)^{(m+3)/2} \frac{Q}{a \, \alpha_m^2} + \frac{2 \, P_o}{a \, \alpha_m^3} + \frac{4 \, P \, \delta}{a^3 \, \alpha_m^3} +$$

$$\frac{P \, \alpha_m}{a \, (\alpha_m^2 + \theta^2)} \, (\frac{P}{b \, G} - h)$$

$$A''_m = P \, (\frac{N_m}{b \, G} + \mu \, \cosh \frac{1}{2} \mu \, \alpha_m \, t_c) - (M_m + h \, N_m)$$

The deflection amplitude $W_m$ can thus be determined from either Equ. (4.50) or (4.51). Knowing $C_m$ and $W_m$, the quantities of interest such as stress distribution, interlayer shear, and deflection can readily be obtained.

Particular cases can now be obtained from the previous general analysis. In specific, solutions for a beam subjected to three loading types are presented next.

### 4.4.2.1 UNIFORM LOAD OF INTENSITY $p_o$

In this case $Q = P = \delta = 0$, and the Equs. (4.50) and (4.51) yield:

$$C_m = \frac{- 2 \, P_o}{a \, \alpha_m^3 \, (M_m + h \, N_m)} \tag{4.53}$$

$$W_m = - C_m \, (\mu \, \cosh \frac{1}{2} \mu \, \alpha_m \, t_c + \frac{N_m}{b \, G}) \tag{4.54}$$

### 4.4.2.2 MIDSPAN CONCENTRATED LOAD

In this case $p_o = P = \delta = 0$, and Equ. (4.52) yields:

$$C_m = \frac{(-1)^{(m+3)/2} \; Q}{a \, \alpha_m^2 \, (M_m + h \, N_m)} \tag{4.55}$$

while $W_m$ is as given by Eus. (4.54).

### 4.4.2.3 BUCKLING LOAD FOR COMPOSITE PANELS

In this case $p_o = Q = 0$, and Equs. (4.51) and (4.52) yield:

$$C_m = \frac{A_m'''}{A_m''} \tag{4.56}$$

$$W_m = \frac{W_m'}{A_m''} \tag{4.57}$$

in which

$$A_m''' = \frac{4 \, P \, \delta}{a^3 \, \alpha_m^3} + \frac{P \, \alpha_m}{a \, (\alpha_m^2 + \theta^2)} \left(\frac{P}{b \, G} - h\right)$$

$$W_m' = \frac{P \, \alpha_m \, A_m''}{G \, a \, b \, (\alpha_m^2 + \theta^2)} - A_m''' \left(\frac{N_m}{b \, G} + \mu \cosh \frac{1}{2} \mu \, \alpha_m \, t_c\right)$$

The deflection w becomes infinitely large when the denominator in the right hand side of Equ. (4.57) becomes zero. Thus an equation for determining the critical condition is [22]:

$$P \left(\frac{N_m}{b \, G} + \mu \cosh \frac{1}{2} \mu \, \alpha_m \, t_c\right) - (M_m + h \, N_m) = 0$$

from which

$$P_{cr} = \frac{M_m + h \, N_m}{(N_m / b \, G) + \mu \cosh \frac{1}{2} \mu \, \alpha_m \, t_c} \tag{4.58}$$

The smallest value of the critical load can be obtained by taking m = 1.

## 4.5 PARAMETRIC STUDY OF BONDING EFFECTS ON THE BEHAVIOR OF COMPOSITE BEAM-COLUMNS

The complexity of the results obtained in the previous section makes it difficult to see the effects of bonding on the response of composite beams. For the purpose of illustrating these effects, a particular beam is considered. The beam has the following properties:

$$a = 20. \text{ in.}$$
$$t_f = 0.04 \text{ in.}$$
$$t_c = 2. \text{ in.}$$
$$b = 1. \text{ in.}$$
$$E_f = 10^7 \text{ psi}$$
$$E_c = 2 \times 10^4 \text{ psi}$$
$$G = 10^4 \text{ psi}$$

The main parameter of interest is the bonding stiffness K. Its effects on the following quantities are shown in Fig. 4.10, 4.11, and 4.14.

$$(w)_{x=a} = w$$

$$(N)_{x=a} = N$$

$$(\sigma_c)_{\substack{x=a \\ z=t_c/2}} = \sigma_c$$

$$(q)_{\substack{x=0 \\ z=t_c/2}} = q$$

in which
- $w$ = the central deflection;
- $N$ = the normal force in the faces at mid span;
- $\sigma_c$ = the maximum normal stress in the core at mid span;
- $q$ = the interlayer shear at the beam end.

It is seen in these figures that the deflection shows greater sensitivity to variation of the K value when the latter is in the lower range, and beyond a certain level of bonding stiffness, the bonding can be practically considered as rigid. For instance, in the case of uniform loading (Fig. 4.10), a change in K value from 3000 to 2000 psi. induces a deflection increase fifteen times greater than that when K changes from 10000 to 9000 psi. It is noted that the level of K for a rigid adhesive varies little with the types of lateral loading, however, it is substantially lower than that in the case of buckling due to axial compressive load.

An increase in the bonding stiffness is accompanied by a decrease in the normal stress of the core. The resulting loss in the resisting moment is compensated by a slight increase in the face normal stress. For example, in Fig. 4.10, a change in K value from 8000 to 9000 psi. induces a decrease in the bending moment contributed by the core by $0.087 p_0$, whereas the increase in the moment contributed by the facings is $0.088 p_0$. The interlayer shear flux, q, is practically independent of the bonding stiffness (Fig. 4.10 and 4.11). The bonding effects on the normal stress, $\sigma_c$, and shear stress, $\tau$, are shown in Figs. 4.12 and 4.13. It is seen that by increasing the K value the normal stress is reduced and the shear stress approaches a linear distribution with constant value.

## 4.6   PRACTICAL FORMULAS FOR COMPOSITE PANELS

To facilitate the application of the present theories, simple formulas are derived from which the maximum deflection, normal force in facings, and interlayer flux can be determined. Three loading types are considered separately: (i) uniformly distributed load, (ii) mid span concentrated load, and (iii) axial compression.

### 4.6.1   UNIFORM LOAD OF INTENSITY $p_0$

The quanitites of primary interest are: the maximum normal force in facings, the maximum interlayer shear flux, and the maximum deflection. These quantities are calculated from Equs. (4.44), (4.39), and (4.45) in conjunction with Equs. (4.53) and (4.54). The results are written in abbreviated form as :

$$N = (N)_{x=a} = \frac{1}{2} p_0 \, t_c \, F_{nu} \qquad (4.59)$$

$$q = (q)_{x=0} = p_0 \, F_{qu} \qquad (4.60)$$

$$w = (w)_{x=a} = \frac{p_0 \, t_c}{2 \, E_c \, b} \, F_{wu} \qquad (4.61)$$

in which

N = the maximum normal force in facings;
q = the maximum interlayer shear flux;
w = the maximum deflection.

$$F_{nu} = \sum_{m=1,3,\ldots}^{\infty} \frac{-16 \, \mu \, a}{\pi^2 \, m^2 \, t_c} \, (-1)^{(m+1)/2} \, F_m$$

$$F_{qu} = \sum_{m=1,3,\ldots}^{\infty} \frac{4 \, \mu}{m \, \pi} \, F_m$$

$$F_{wu} = \sum_{m=1,3,\ldots}^{\infty} \frac{-64 \, (-1)^{(m+1)/2} \, \mu^2 \, a^2}{\pi^3 \, m^3 \, t_c^2} \, F_m \left[ \left( \xi + \frac{m^2 \, \pi^2 \, t_c^2 \, \phi}{16 \, a^2} \right) \right.$$
$$\left. \coth \frac{m \, \pi \, t_c \, \mu}{4 \, a} + \frac{m \, \pi \, t_c \, \mu}{4 \, a} \right]$$

where

$$\xi = \frac{E_c \, t_c}{2 \, E_f \, t_f}$$

$$\phi = 2 \, \frac{E_c \, b}{K \, t_c}$$

$$F_m = \frac{\sinh (m \, \pi \, t_c \, \mu / 4 \, a)}{F_m'}$$

$$F_m' = \left( \xi + \frac{m^2 \, \pi^2 \, t_c^2 \, \phi}{16 \, a^2} \right) \left( 2 \cosh \frac{m \, \pi \, t_c \, \mu}{4 \, a} - \right.$$
$$\left. \frac{8 \, a}{m \, \pi \, t_c \, \mu} \sinh \frac{m \, \pi \, t_c \, \mu}{4 \, a} \right) + \frac{m \, \pi \, t_c \, \mu}{2 \, a} \sinh \frac{m \, \pi \, t_c \, \mu}{4 \, a}$$

Numerical values for $F_{nu}$, $F_{qu}$, and $F_{wu}$ are obtained for the following range of $\xi$, $\bar{I}$, and $(2a/t_c)$:

(i)     $\xi$         varies from 0.05 to 0.30.
(ii)    $\bar{I}$         varies from 0.0 to 10.
(iii)   $2a/t_c$     varies from 20. to 50.

The results are presented in Figs. 4.15 to 4.18.


## 4.6.2   MID SPAN CONCENTRATED LOAD

In this case, the quantities N, q, and w are obtained from Equs. (4.44), (4.39), and (4.45) in conjunction with Equs. (4.55) and (4.54) as:

$$N = QF_{nc} \tag{4.62}$$
$$q = (Q/a)F_{qc} \tag{4.63}$$
$$w = (Q/E_c b)F_{wc} \tag{4.64}$$

in which

$$F_{nc} = \sum_{m=1,3,\ldots}^{\infty} \frac{2\mu}{m\pi} F_m$$

$$F_{qc} = \sum_{m=1,3,\ldots}^{\infty} -(-1)^{(m+1)/2} \mu F_m$$

$$F_{wc} = \sum_{m=1,3,\ldots}^{\infty} \frac{-8(-1)^{(m+1)/2}\mu^2 a}{m^2 \pi^2 t_c} F_m \left[ (\xi + \frac{m^2 \pi^2 t_c^2 \phi}{16 a^2}) \right.$$

$$\left. \coth \frac{m\pi t_c \mu}{4a} + \frac{m\pi t_c \mu}{4a} \right]$$

Numerical values for $F_{nc}$, $F_{qc}$, and $F_{wc}$ are obtained for the same range of $\xi$, $\bar{I}$, and $(2a/t_c)$ considered in the previous section. The results are presented in Figs. 4.19 to 4.22.

### 4.6.3   BUCKLING LOAD FOR COMPOSITE PANELS

The buckling load in this case is determined from Equ. (4.58) as:

$$P_{cr} = \frac{E_c \, t_c \, b}{2} \, F_p \qquad\qquad (4.65)$$

in which
$$F_p = F_p{}'/F_p{}''$$

where

$$F_p{}' = (\xi + \frac{m^2 \, t_c^2 \, \Phi}{16 \, a^2}) \, (2 \cosh \frac{\pi \, t_c \, \mu}{4 \, a} - $$

$$\frac{8 \, a}{\pi \, t_c \, \mu} \sinh \frac{\pi \, t_c \, \mu}{4 \, a}) + \frac{\pi \, t_c \, \mu}{2 \, a} \sinh \frac{\pi \, t_c \, \mu}{4 \, a}$$

$$F_p{}'' = \mu^2 \, [\frac{\pi \, t_c \, \mu}{4 \, a} \sinh \frac{\pi \, t_c \, \mu}{4 \, a} + (\xi + \frac{\pi^2 \, t_c^2 \, \Phi}{16 \, a^2}) \cosh \frac{\pi \, t_c \, \mu}{4 \, a}]$$

Numerical values for $F_p$ can be obtained from Figs. 4.23 and 4.24 for the same range of $\xi$, $\ddot{I}$, and $(2a/t_c)$ considered in section 4.6.1.

### 4.7   REFERENCES

[1]   Allen, H.G., Analysis and Design of Structural Sandwich Panels, First edition, Pergamon Press Ltd., 1969.

[2]   Amana, E.J., and Booth, L.G., "Theoretical and Experimental Studies on Nailed and Glued Plywood Stressed-Skin Components:  Part I.  Theoretical Study", J. Inst. of Wood Science, Vol. 4, No. 1, 1967.

[3] Chang, C.C., and Ebcioglu, I.K., "Thermoelastic Behaviour of a Simply Supported Sandwich Panel Under Large Temperature Gradient and Edge Compression", J. of the Aerospace Science, Vol. 28, No.6, June, 196.

[4] Feldman, D., "Private Communication", Centre for Building Studies, Concordia University, 1979.

[5] Goodier, J.N., and Hsu, C.S., "Nonsinusoidal Buckling Modes of Sandwich Plates", J. Aero. Sci., Vol. 21, No. 8, August, 1954.

[6] Goodman, J.R., and Popov, E.P., "Layered Beam Systems with Inter-Layer Slip", J. Struct. Div., ASCE, Vol. 94, No. ST11, November, 1968.

[7] Goodman, J.R., "Layered Wood Systems with Interlayer Slip", Wood Science, Vol. 1, No. 3, 1969.

[8] Goodman, J.R., Vanderbilt, M.D., Criswell, M.E., and Bodig, J., "Composite and Two-Way Action in Wood Joist Floor Systems", Wood Science, Vol. 7, No. 1, 1974.

[9] Harris, B.J., and Nordby, G.M., "Local Failure of Plastic-Foam Core Sandwich Panels", J. Struct. Div., ASCE, Vol. 95, No. ST4, April, 1969.

[10] Hoff, N.J., and Mautner, S.E., "The Buckling of Sandwich-Type Panels", Journal of the Aeronautical Sciences, Vol. 12, No. 3, July, 1945.

[11] Hussein, R., "Thermal Stress in Flat Metal-Faced Sandwich Panel", M. Eng. Thesis, Centre for Building Studies, Concordia University, Montreal, 1978.

[12] Isakson, G., and Levy, A., "Finite Element Analysis of Inter-Laminar Shear in Fibrous Composites", J. Composite Materials, Vol. 5, April, 1971.

[13] Kuenzi, E.W., and Wilkinson, T.L., "Composite Beams-Effect of Adhesive or Fastener Rigidity", U.S. Forest Products Laboratory, Research Paper No. FPL 152, January, 1971.

[14] Marsh, C., Private Communication, Center for Building Studies, Concordia University, 1981.

[15] McGlenn, J., and Hartz, B., "Finite Element Analysis of Plywood Plates", J. Struct. Div., ASCE, Vol. 94, No. ST2, February, 1968.

[16] Pagano, N.J., "Exact Solutions for Composite Laminates in Cylindrical Bending", J. Composite Materials, Vol. 3, July, 1969.

[17] Pagano, N.J., "Exact Solutions for Rectangular Bidirectional Composites and Sandwich Plates", J. Composite Materials, Vol. 4, January, 1970.

[18] Plantema, F.J., Sandwich Construction - The Bending and Buckling of Sandwich Beams, Plates and Shells, John Wiley and Sons, Inc., 1966.

[19] Puppo, A.H., and Evensen, H.A., Interlaminar Shear in Laminated Composites under Generalized Plane Stress", J. Composite Materials, Vol. 4, April, 1970.

[20] Thompson, E.G., Goodman, J.R., and Vanderbilt, M.D., "Finite Element Analysis of Layered Wood Systems", J. Struct. Div., ASCE, Vol. 101, No. ST12, December, 1975.

[21] Thompson, E.G., Vanderbilt, M.D., and Goodman, J.R., "FEAFLO - A Program for the Analysis of Layered Wood Systems", Computers and Structures, Vol. 7, 1977.

[22] Timoshenko, S.P., and Gere, J.M., Theory of Elastic Stability, Second edition, McGraw-Hill Book Company, Inc., 1961.

[23] Vanderbilt, M.D., Goodman, J.R., and Criswell, M.E., "Service and Overload Behaviour of Wood Joist Floor Systems", J. Struct. Div., ASCE, Vol. 100, No. St1, January, 1974.

[24] Whitney, J.M., "Bending-Extensional Coupling in Laminated Plates under Transverse Loading", J. Composite Materials, Vol. 3, January, 1969.

[25] Whitney, J.M., and Leissa, A.W., "Analysis of Heterogeneous Anisotropic Plates", Transactions of the ASME, J. of Applied Mechanics, Vol. 28, June, 1969.

[26] Yusuff, S., "Theory of Wrinkling in Sandwich Construction", J. of the Royal Aeronautical Society, Vol. 59, January, 1955.

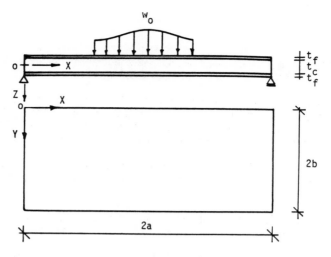

Fig. 4.1   COMPOSITE PLATE

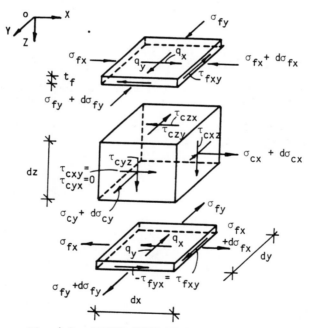

Fig. 4.2   STRESS STATE IN COMPOSITE ELEMENT

128

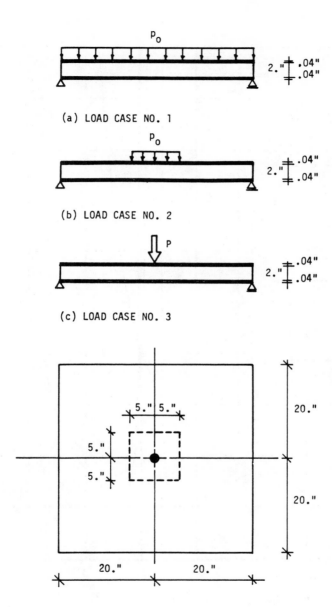

(a) LOAD CASE NO. 1

(b) LOAD CASE NO. 2

(c) LOAD CASE NO. 3

Fig. 4.3    COMPOSITE PLATE IN THE PARAMETRIC STUDY

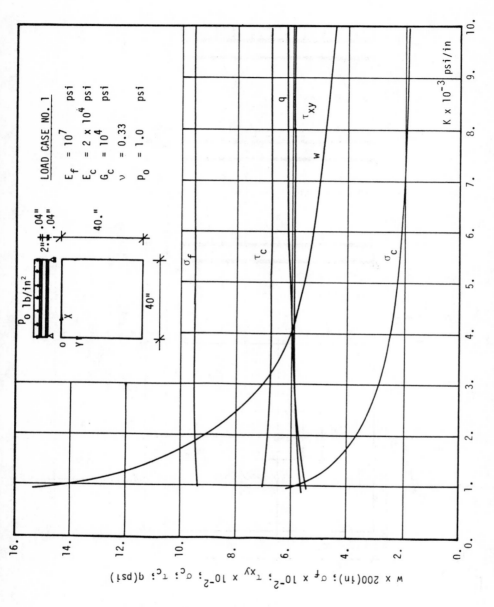

Fig. 4.4 BONDING EFFECTS ON THE DEFLECTION AND STRESSES IN THE COMPOSITE PLATE IN Fig. 4.3

**130**

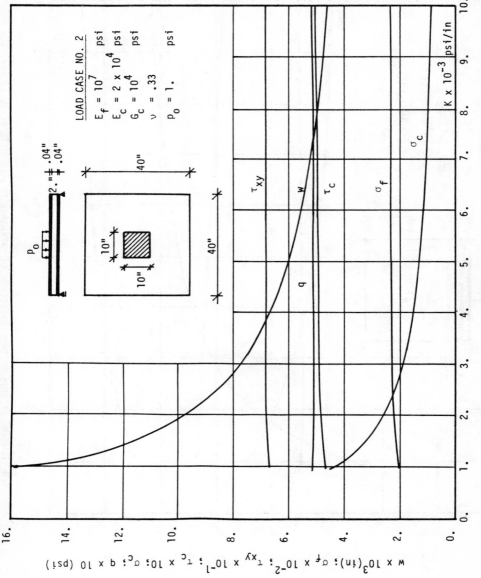

Fig. 4.5  BONDING EFFECTS ON THE DEFLECTION AND STRESSES IN THE COMPOSITE PLATE IN Fig. 4.3

**131**

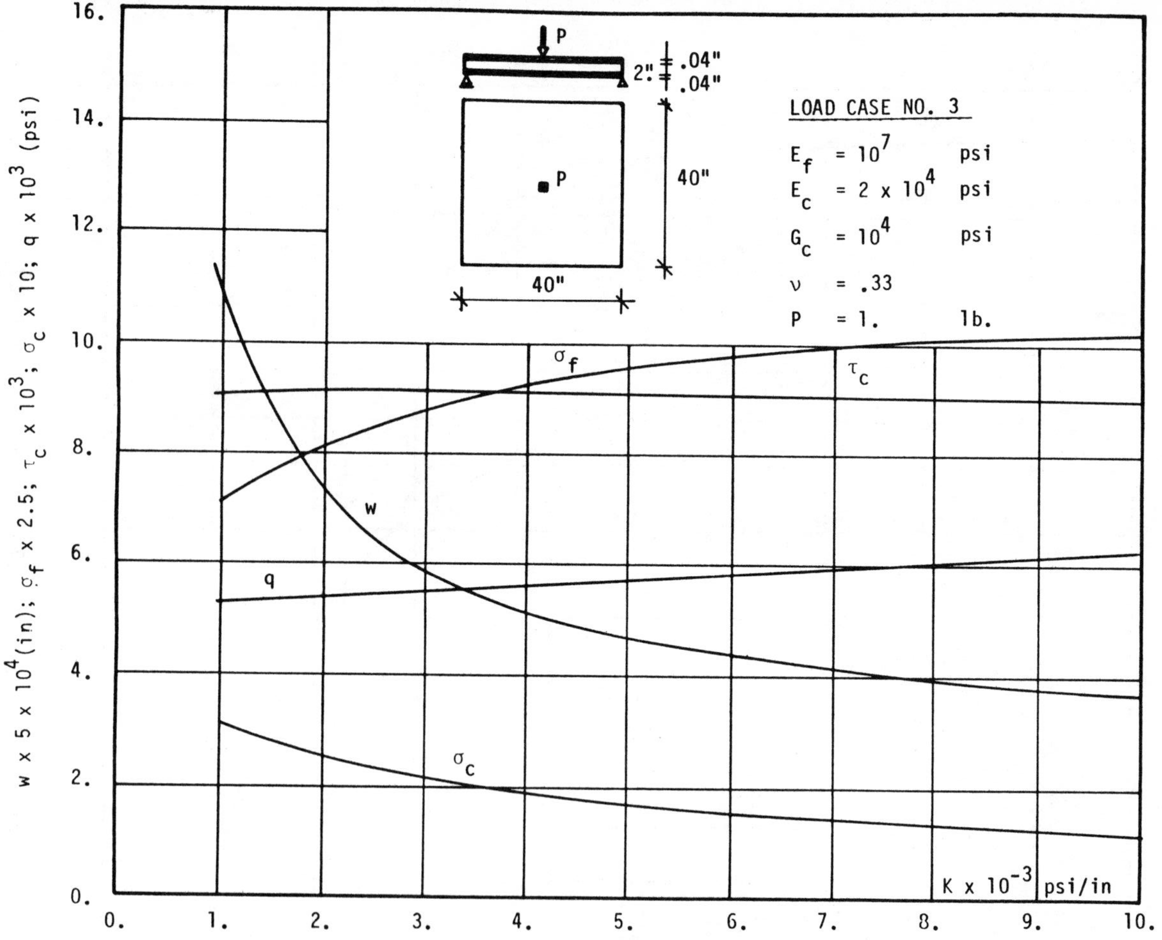

Fig. 4.6 BONDING EFFECTS ON THE DEFLECTION AND STRESSES IN COMPOSITE PLATE IN Fig. 4.3

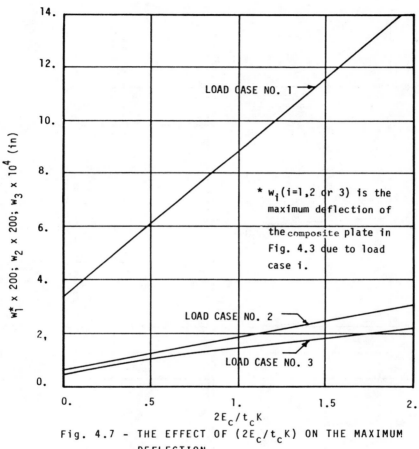

Fig. 4.7 - THE EFFECT OF $(2E_c/t_c K)$ ON THE MAXIMUM DEFLECTION

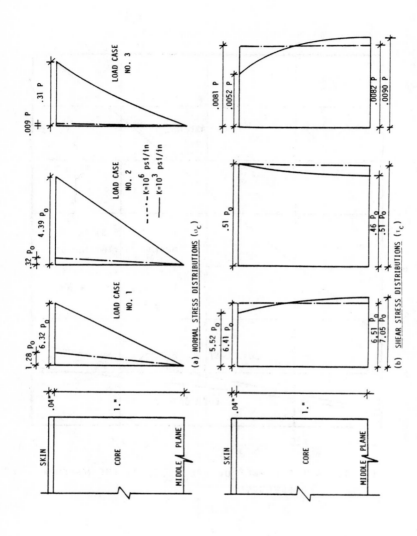

Fig. 4.8  BONDING EFFECTS ON STRESS DISTRIBUTIONS IN THE INNER LAMINA OF THE PLATE IN Fig. 4.3

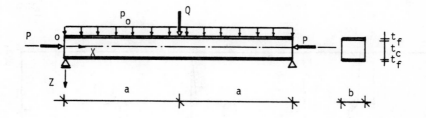

(a) LOADS AND GEOMETRY

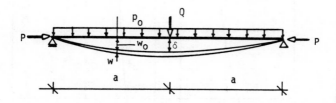

(b) DEFLECTION AND INITIAL IMPERFECTION

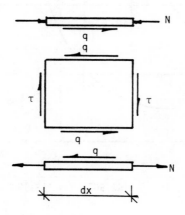

(c) INTERNAL STRESSES

Fig. 4.9 COMPOSITE BEAM-COLUMN

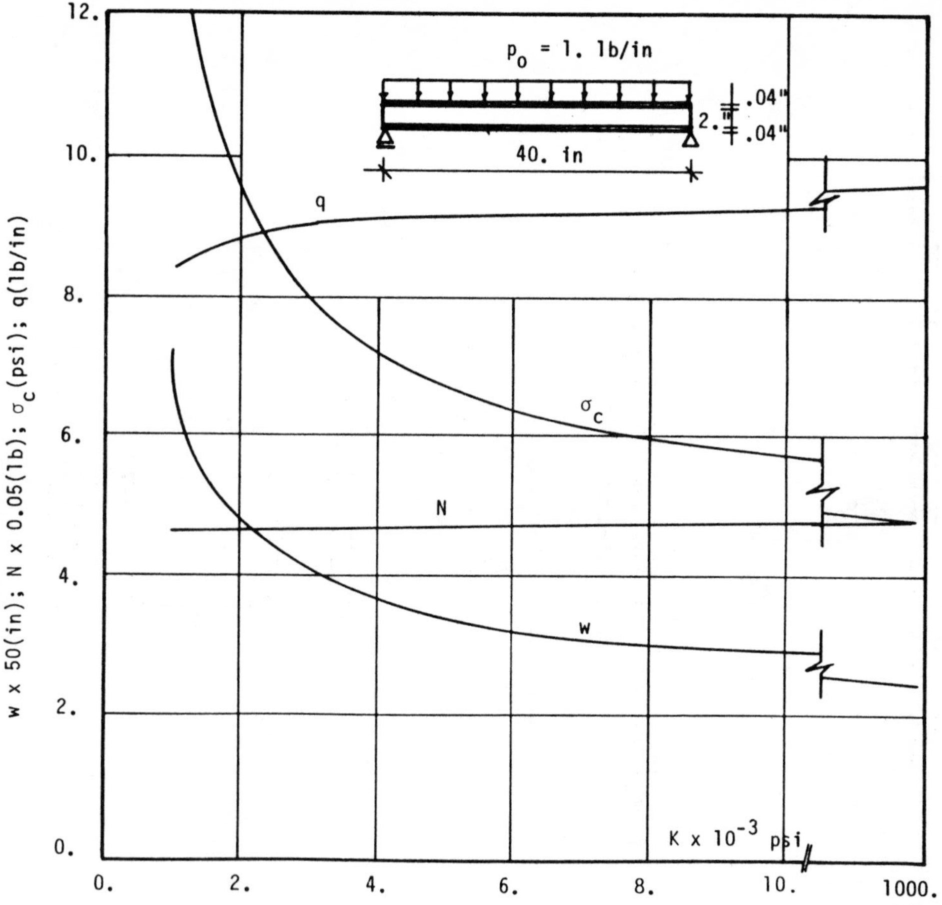

Fig. 4.10 BONDING EFFECTS ON COMPOSITE BEAM BEHAVIOR

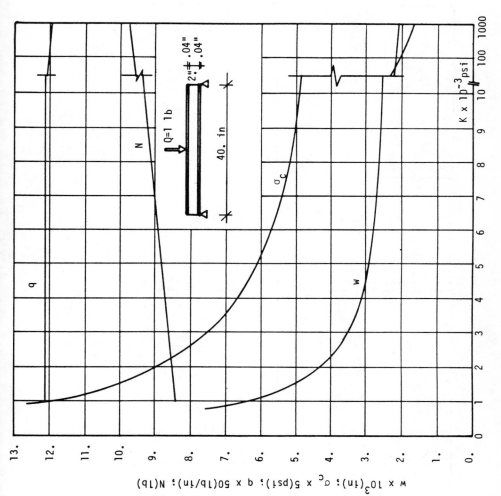

Fig. 4.11 BONDING EFFECTS ON COMPOSITE BEAM BEHAVIOR

**137**

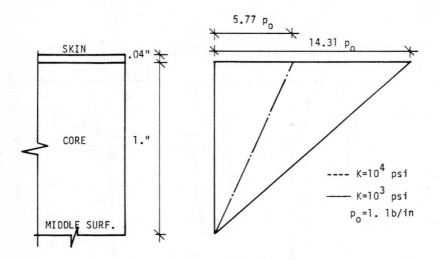

(a) NORMAL STRESS DISTRIBUTION ($\sigma_c$)

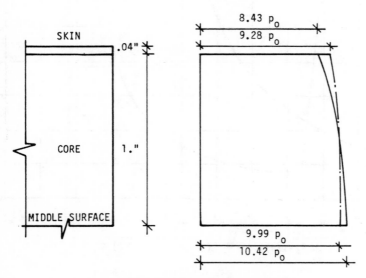

(b) SHEAR STRESS DISTRIBUTION ($\tau$)

Fig. 4.12   BONDING EFFECTS ON STRESS DISTRIBUTION IN
THE INNER LAMINA

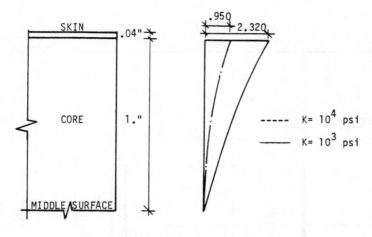

(a) <u>NORMAL STRESS DISTRIBUTION ($\sigma_c$)</u>

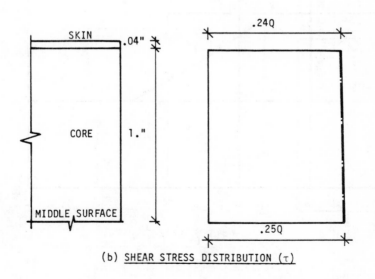

(b) <u>SHEAR STRESS DISTRIBUTION ($\tau$)</u>

Fig. 4.13   BONDING EFFECT ON STRESS DISTRIBUTIONS IN
            THE INNER LAMINA

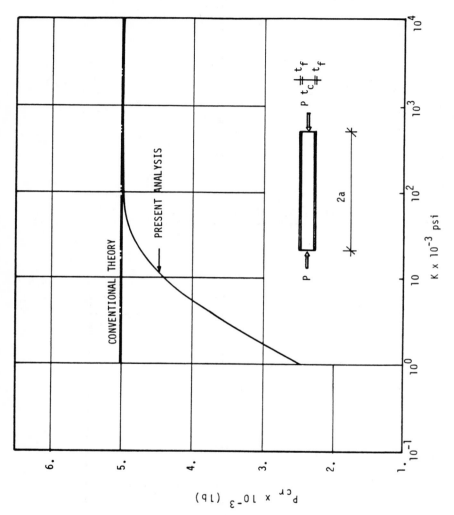

Fig. 4.14  BONDING EFFECT ON THE CRITICAL LOAD OF A COMPOSITE PANEL

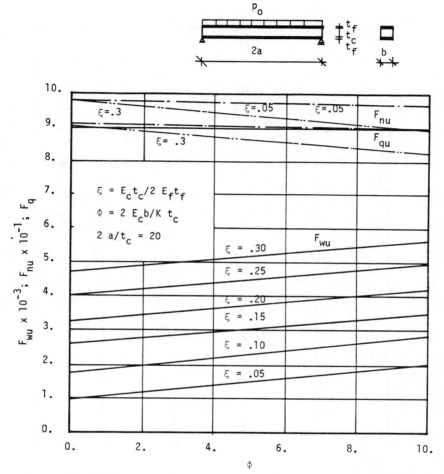

Fig. 4.15 - NUMERICAL VALUES FOR FACTORS IN EQUS.(4.44),(4.45),
AND (4.46)

141

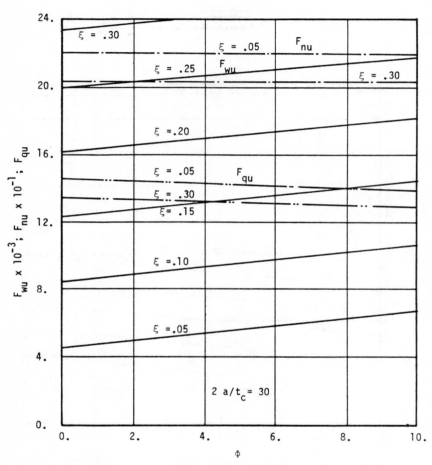

Fig. 4.16 - NUMERICAL VALUES FOR FACTORS IN EQUS.(4.44), (4.45),
AND (4.46)

142

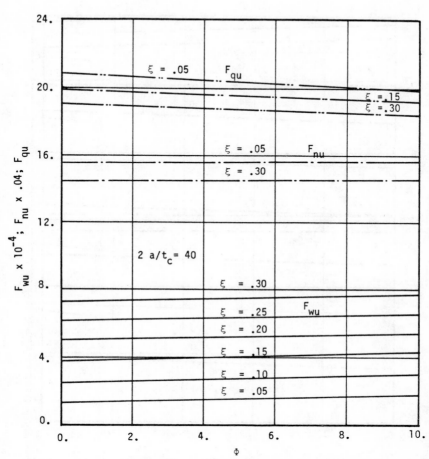

Fig. 4.17 - NUMERICAL VALUES FOR FACTORS IN EQUS.(4.44), (4.45), AND (4.46)

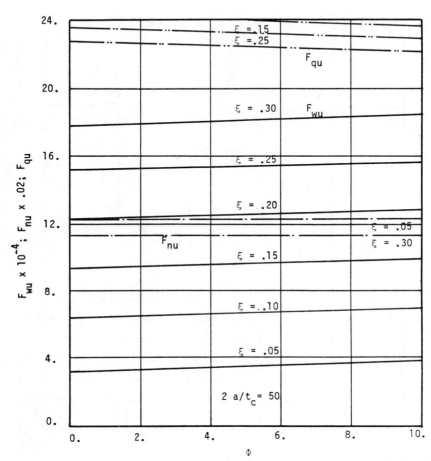

Fig. 4.18 - NUMERICAL VALUES FOR FACTORS IN EQUS.(4.44), (4.45), AND (4.46)

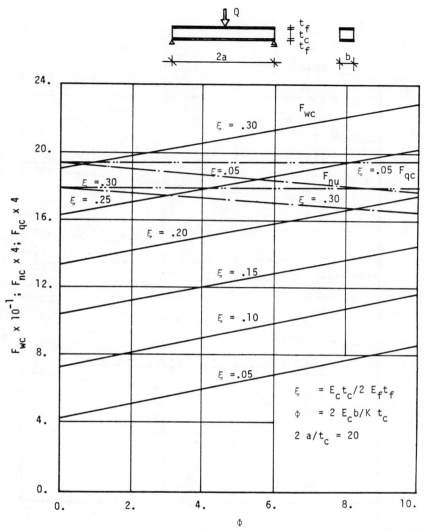

Fig. 4.19 - NUMERICAL VALUES FOR FACTORS IN EQUS.(4.47),(4.48),
AND (4.49)

145

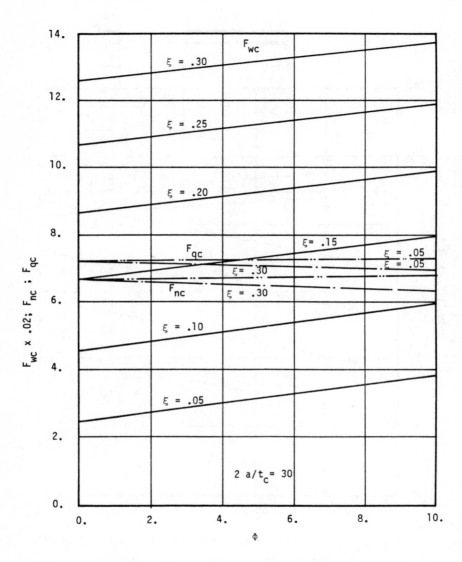

Fig. 4.20 - NUMERICAL VALUES FOR FACTORS IN EQUS.(4.47), (4.48),
AND (4.49)

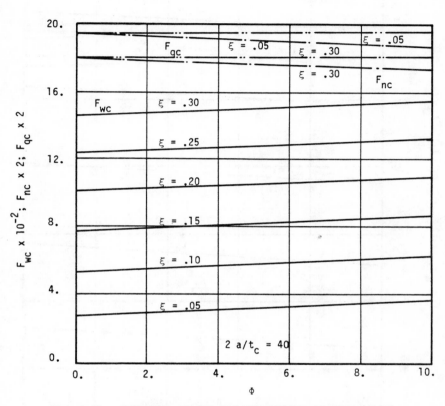

Fig. 4.21 - NUMERICAL VALUES FOR THE FACTORS IN EQUS.(4.47), (4.48), AND (4.49)

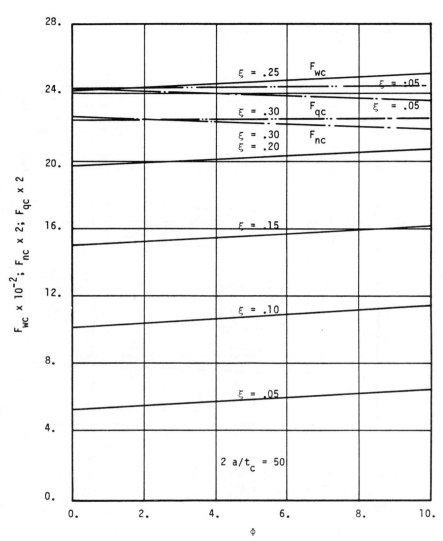

Fig. 4.22 - NUMERICAL VALUES FOR FACTORS IN EQUS.(4.47), (4.48),
AND (4.49)

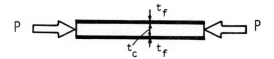

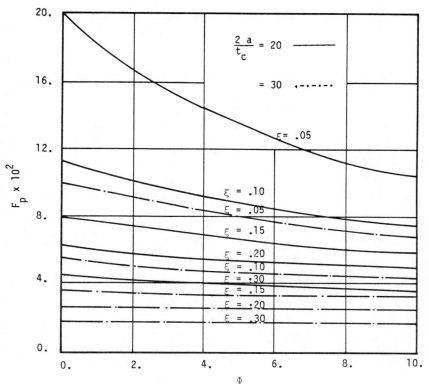

Fig. 4.23 - NUMERICAL VALUES FOR FACTOR IN EQU.(4.50)

149

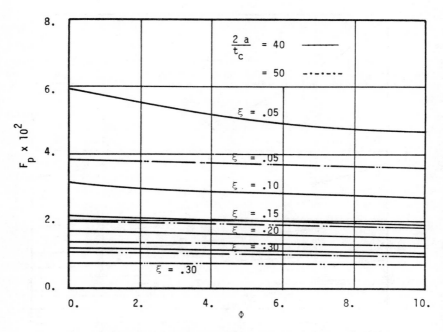

Fig. 4.24 - NUMERICAL VALUES FOR FACTOR IN EQU. (4.50)

# Instability of Composite Panels

## 5.1 INTRODUCTION

The elastic instability problem of composite panels is investigated in this chapter. The two major failure modes, in which instability may manifest itself are: overall buckling and local instability.

Overall buckling occurs when a panel becomes elastically unstable under the application of inplane loads with the buckling mode characterized by long waves. This problem has been extensively investigated by many investigators both theoretically and analytically. Analytic solutions for the critical loads, in correlation with experimental results are found in References [1, 9, 10, 11, 12, 14]. Existing work covers a large variety of loading types, boundary conditions, and panel fabrications. A comprehensive coverage of this topic is found in References [3, 19]. For the present chapter, only local instability of panels is considered.

Local instability of panels may be classified into three categories: dimpling, wrinkling, and crimping. Dimpling occurs in composite construction with a honeycomb core, where in the region above the cell, the face buckles in a plate-like fashion with the cell walls acting as edge supports. Failure by wrinkling occurs in the form of short wave lengths in the facings, and involves straining of the core material. Such failure mode could be symmetric or antisymmetric with respect to the middle plane of the whole panel. Finally, crimping is a shear failure mode.

In this chapter, analytical solutions for the three local failure modes are presented. By applying the finite difference method, a numerical value for the dimpling coefficient is found. Comparison is made with existing semi-empirically derived coefficient. Analytic solutions for the wrinkling stress and failing stresses of the core material are also developed. By adopting the available theory by Hoff and Mautner [9], a solution is found to define the transition state between the symmetric and antisymmetric wrinkling modes. Although a solution for the crimping load has been developed, the approach presented here is believed to be simpler and suitable for practical purposes.

Previous work related to the instability phenomena of dimpling was carried out at the Forest Products Laboratory [15] where paper honeycomb panels and panels having a solid spruce core through which circular holes were drilled to simulate a core cell were used in the tests. Based on the experimental results, a semi-empirical formula for the dimpling stress was developed [15]. With regard to the wrinkling load and failing stress of core materials two approaches were used in the past. Hoff and Mautner [9] made use of the principle of strain energy with a linear distribution of the transverse displacement through the core. On the other hand, Chong and Hartsock [2]; Harris et al. [7, 8]; and Yusuff [21,22] considered the faces as plates on an elastic foundation, for which an equivalent foundation modulus was obtained by equating the strain energy stored in the foundation to the sum of the extensional and shear strain energies stored in the core. In all the investigations, experimental works were conducted. Hoff and Mautner tested fifty-one flat rectangular specimens in edgewise compression, and observed both modes of the wrinkling.

## 5.2   DIMPLING OF COMPOSITE PANELS

The critical stress causing elastic buckling of a homogeneous plate subjected to compressive forces can be put in the following form [18]:

$$\sigma_{cr} = \frac{K \pi^2 E_f}{12 (1 - v_f^2)} (\frac{t_f}{S})^2 \tag{5.1}$$

in which

K    =    coefficient which depends on the plate boundary conditions, and loading type;

$E_f$    =    elastic modulus of the plate material;

$v_f$    =    Poisson's ratio of the plate material;

$t_f$    =    the plate thickness;

S    =    the plate width.

Equation (5.1) can be written as:

$$(\frac{t_f}{S})^2 = \frac{\sigma_{cr} (1 - v_f^2)}{K_d E_f} \tag{5.2}$$

in which

$$K_d = \frac{\pi^2 K}{12}$$

Since dimpling occurs in a plate-like fashion, where in the region above a cell, in the core, the face buckles with the cell walls acting as edge supports, Equation (5.2) may be used to estimate the dimpling stress. The symbol S is now interpreted to be the diameter of the largest circle that can be inscribed within a cell of the core [15] (Fig. 5.1(a)); $E_f$, $v_f$, $t_f$ are the facing properties. With this procedure in conjunction with experimental results (Fig. 5.1(b)), the Forest Products Laboratory [15] has obtained a coefficient of 2 for $K_d$.

As an attempt to give some theoretical justification to this semi-empirical coefficient, the finite difference method is now used to determine the buckling stress for a plate supported by honeycombed core cells. For this purpose, consider the plate in Figure 5.2. The governing differential equation for the buckling of a thin plate is [16, 17, 18]:

$$\nabla^4 w + \frac{P_x}{D} \frac{\partial^2 w}{\partial x^2} = 0 \qquad (5.3)$$

in which

$$P_x = \text{the applied compressive force, per unit length, of the plate in X- direction (Fig. 5.2)}$$

$$\nabla^4 = \frac{\partial^4}{\partial x^4} + 2 \frac{\partial^4}{\partial x^2 \partial y^2} + \frac{\partial^4}{\partial y^4}$$

$$D = \text{the flexural rigidity of the plate}$$

$$= \frac{E_f \, t_f^3}{12 \, (1 - v_f^2)}$$

$$w = \text{plate deflection.}$$

By applying the ordinary finite difference stencil [16, 17] to the finite difference mesh points (Fig. 5.2), the resulting system of linear simultaneous equations is:

$$[A]\{w\} = \alpha[B]\{w\}$$

or

$$[C]\{w\} = \alpha\{w\} \tag{5.4}$$

in which

$\alpha$ = $\quad p_x \lambda^2/D$

where $\lambda$ is the mesh width, $\lambda = a/3$ and a is the side-length of a square within which the cell in Fig. 5.2 is inscribed;

$\{w\}$ = the vector of the deflection ordinates at the finite difference mesh points;

$[C]$ = $[B]^{-1}[A]$, where [A] and [B] are coefficient matrices obtained as

$$[A] = \begin{array}{cccc} 18 & -8 & -8 & 2 \\ & 18 & 2 & -8 \\ \text{Sym.} & & 18 & -8 \\ & & & 18 \end{array}$$

$$[B] = \begin{array}{cccc} 2 & -1 & 0 & 0 \\ & 2 & 0 & 0 \\ \text{Sym.} & & 2 & -1 \\ & & & 2 \end{array}$$

Hence, the governing equation (5.3) is reduced to an eigenvalue problem in Equation (5.4). Of the four eigenvalues obtained from a computerized solution [6], the lowest one is

$\alpha = 4.$

Thus, from the definition of $\alpha$, the critical stress is found as

$$\sigma_{cr} = 2.25 \frac{E_f}{(1 - \nu_f^2)} \left(\frac{t_f}{S}\right)^2 \tag{5.5}$$

The coefficient for $K_d$ is then 2.25, which is slightly higher than the value recommended by the Forest Products Laboratory.

For the case when the facings are subjected to biaxial compression, the interaction formula developed by Timoshenko [18] for homogeneous plates may be used

$$\sigma_x + \sigma_y = \sigma_{cr} \tag{5.6}$$

in which

$\sigma_x, \sigma_y$    =    the applied compressive stresses in X and Y directions, respectively.

This formula was recommended by Sullins [15] for dimpling analysis. When one of the applied stresses is tensile, a conservative solution could be obtained by assuming that the compressive stress is acting alone. For more general loading types, the principal stresses should be used in Equation (5.6) [15] in the place of $\sigma_x$ and $\sigma_y$.

## 5.3 WRINKLING OF COMPOSITE PANELS

As has been mentioned, wrinkling is characterized by its short waves involving bending of the skins and compression or elongation of the core material in the transverse direction. This type of local failure occurs when the core thickness is such that the overall buckling is not likely to happen.

In the following, formulas to define the transition state between the symmetric and antisymmetric wrinklings are presented. In addition, solutions for the symmetrical wrinkling stress, and the failing stresses of the core material are developed.

### 5.3.1 TRANSITION STATE BETWEEN THE TWO WRINKLING MODES

In the symmetrical wrinkling mode (Fig. 5.3(a)), the core material is called upon to experience normal and shear stresses within a marginal zone adjacent to the faces. In the middle part of the core, little deformation takes place. By applying the strain energy method, Hoff and Mautner obtained the wrinkling stress for this failure mode as [9]:

$$\sigma_{crs} = 0.817 \left( \frac{E_f \, E_c \, t_f}{t_c} \right)^{1/2} + 0.166 \, \frac{G_c \, t_c}{t_f} \quad \text{when } w = \frac{t_c}{2} \quad (5.7)$$

$$= 0.91 \, (E_f \, E_c \, G_c)^{1/3} \qquad \text{when } w < \frac{t_c}{2} \quad (5.8)$$

in which

$\sigma_{crs}$    =    the symmetrical wrinkling stress;

w    =    the width of the marginal zone (Fig. 5.3(a))

      =    $1.44 \, t_f (E_f/E_c)^{1/3}$;

$t_c$ = the core thickness.

When the core is neither thin enough to cause overall buckling, nor thick enough to cause symmetric wrinkling, the possible wrinkling mode is a skew ripple as shown in Fig. 5.3(b). By using the strain energy method, the wrinkling stress in this case was obtained as [9]:

$$\sigma_{cra} = 0.59 \left( \frac{E_f E_c t_f}{t_c} \right)^{1/2} + 0.387 \frac{G_c t_c}{t_f} \quad \text{when } w = \frac{t_c}{2} \quad (5.9)$$

$$= 0.51 \left( E_f E_c G_c \right)^{1/3} + 0.33 \frac{G_c t_c}{t_f} \quad \text{when } w < \frac{t_c}{2} \quad (5.10)$$

in which

$\sigma_{cra}$ = the antisymmetrical wrinkling stress;

$w$ = $2.38 \, t_f \, (E_f/E_c)^{1/3}$.

By equating the right hand side of Equs. (5.8) and (5.10), and of Equs. (5.7) and (5.9), it is found that:

$$\frac{t_c}{t_f} = 1.924 \, (E_f/E_c)^{1/3} \quad \text{when } w < \frac{t_c}{2} \quad (5.11)$$

and

$$\frac{t_c}{t_f} = 1.616 \, (E_f/E_c)^{1/3} \quad \text{when } w = \frac{t_c}{2} \quad (5.12)$$

Equs. (5.11) and (5.12) can be used to predict the correct mode of wrinkling. If the ratio of the core to face thicknesses is less than that given by the expression (5.11) or (5.12) an antisymmetric wrinkling mode is to be expected.

### 5.3.2  ANALYTIC  SOLUTION  FOR  THE  SYMMETRICAL  WRINKLING STRESS

The problem of symmetrical wrinkling of panels was studied by many investigators [2, 7, 8, 9, 21, 22]. In all these studies, a linear distribution of the transverse displacement through the core was considered, and the faces were treated as plates on elastic foundation.

In this section, an analytic solution for the symmetrical wrinkling stress is obtained by using an elasticity approach. The assumptions commonly accepted for this type of analysis are:
(1) The inplane stresses in the core are neglected. That is, with X-Y axes in the plane of a plate, and a Z axis perpendicular to it:

$$\sigma_{cx} = \sigma_{cy} = \tau_{cxy} = 0$$

in which

$\sigma, \tau$ = normal and shear stresses, respectively;
c = subscript denoting the core.

Thus, the relevant deformations in the core are in the transverse direction and shear deformations in XZ and YZ planes.
(2) The wrinkling consists of a plane deformation. Thus, if a panel is compressed in X-direction (Fig. 5.3(a)), the lateral deflection w is independent of y [4, 7, 8, 9, 12, 21, 22].
(3) The core can be treated as a semi-infinite medium in which the displacement decreases exponentially with maximum value at the interface with the skin [5, 12, 20, 21].
(4) The faces are thin in comparison with the core thickness. This implies that the deflection of each face is identified with the displacement of the core at its surface.
(5) The effect of Poisson's ratio of the core material is neglected [4, 8, 9, 13, 21].

Consider now an element of the core in Fig. 3.5(a). The equilibrium equations of such an element are:

$\partial\tau_{zx}/\partial z$ = 0     in X direction
$\partial\tau_{zy}/\partial z$ = 0     in Y direction
$\partial\tau_{xz}/\partial x + \partial\tau_{yz}/\partial y + \partial\sigma_z/\partial z$ = 0     in Z direction

With u, v, w as the core displacement components in X-, Y-, and Z- directions respectively, the stress-strain relations are:

$$\tau_{xz} \doteq G_c \left(\frac{\partial w}{\partial x} + \frac{\partial u}{\partial z}\right)$$

$$\tau_{yz} \doteq G_{cyz} \left(\frac{\partial w}{\partial y} + \frac{\partial v}{\partial z}\right)$$

$$= G_{cyz} \frac{\partial v}{\partial z} \qquad \text{since w is independent of y}$$

$$\sigma_z = E_c \frac{\partial w}{\partial z} \qquad \text{since the effect of Poisson's ratio of}$$

the core material is neglected

and the third equilibrium equation becomes:

$$G_c \frac{\partial^2 w}{\partial x^2} + E_c \frac{\partial^2 w}{\partial z^2} = 0 \qquad (5.13)$$

Because of symmetry about the middle plane of the core (Fig. 5.3(a)), only the left half is considered here, and a solution for Equ. (5.13) is taken as [12]:

$$w = w_o' e^{kz} \sin(\pi x/L)$$

in which

$$L = \text{the wave length;}$$
$$w_o' = \text{the deflection at } x = L/2 \text{ and } z = 0.$$

Substituting of this equation into Equ. (5.13) yields:

$$K = \frac{\pi}{L} \left(\frac{G_c}{E_c}\right)^{1/2}$$

Consequently,

$$w_s = (w)_{z=0} = w_o' \sin(\pi x/L)$$
$$\sigma_s = (\sigma_z)_{z=0} = E_c K w_o' \sin(\pi x/L)$$

The equilibrium equations of a face element have the form of beam-column equations [18]:

$$\frac{dM}{dx} - Q = 0$$

$$-\frac{dQ}{dx} + P \frac{d^2 w_s}{dx^2} + \sigma_s = 0$$

$$M = -D \frac{d^2 w_s}{dx^2}$$

in which

$M, Q$ = the bending moment and shearing force per unit width of the face;
$P$ = the axial force on the face, Fig. 5.3(a);
$\sigma_s$ = the interlayer normal stress;
$D$ = flexural rigidity of the face
= $(1/12) E_f t_f^3$.

By eliminating M and Q in the equilibrium equations, the governing equation for the face deflection is obtained as:

$$D \frac{d^4 w_s}{dx^4} + P \frac{d^2 w_s}{dx^2} = -\sigma_s \qquad (5.14)$$

Substituting the expressions for $w_s$ and $\sigma_s$ into this equation and rearranging the terms gives:

$$P = D \left(\frac{\pi}{L}\right)^2 + E_c K \left(\frac{L}{\pi}\right)^2$$

The above equation represents the axial load required to keep a face in its wrinkling pattern. The critical wave length is the one which makes P a minimum:

$$dP/dL = 0$$

This yields

consequently,

$$P_{cr} = 0.825t_f (E_fE_cG_c)^{1/3}$$

and the critical stress is

$$\sigma_{cr} = 0.825(E_fE_cG_c)^{1/3}$$

In a more general form, Equ. (5.16) can be written as:

$$\sigma_{cr} = K_w(E_fE_cG_c)^{1/3}$$

where $K_w$ is called the wrinkling coefficient and has been found to be equal to 0.961, 0.91, 0.85, and 0.78 respectively in the investigations carried out by Yusuff [21, 22], Hoff [9], Williams [20], and Gough [5]. In practice, an experimental value of 0.5 is commonly used for K [9] (Fig. 5.4).

### 5.3.3  EFFECT OF INITIAL IMPERFECTION ON WRINKLING PHENOMENA

In the foregoing derivations of wrinkling stress, a perfectly flat panel is assumed. To study the effect of an initial imperfection in the form:

$$w_o = w_o{}^" \sin(\pi x/L)$$

in which
$\quad w_o \quad = \quad$ the initial shape of an irregularity in the face;
$\quad w_o{}^" \quad = \quad$ the initial amplitude at $x = L/2$.

on the wrinkling phenomena, the method of equivalent lateral load, as was used by Timoshenko [18], is adopted. According to this method, the governing equation (5.14) becomes:

$$D \frac{d^4w_a}{dx^4} + P \frac{d^2w_a}{dx^2} = -P \frac{d^2w_o}{dx^2} \qquad (5.17)$$

in which

$w_a$ = an additional deflection to be added to the initial deflection when the panel is submitted to the action of the compressive force P.

The deflection $w_a$ is assumed in the form:

$$w_a = w_o'''\sin(\pi x/L)$$

in which

$w_o'''$ = the amplitude of the additional deflection at $x = L/2$

It follows from Equ. (5.17) that:

$$w_o''' = \frac{\alpha\, w_o''}{1 - \alpha} \qquad (5.18)$$

in which

$$\alpha = \frac{P\,L^2}{\pi^2\, E_f\, I_f} = \frac{P}{P_m}$$

$$I = \frac{t_f^3}{12}$$

$$P_m = \frac{\pi^2\, E_f\, I_f}{L^2}$$

Consequently, the total deflection is:

$$w = w_o + w_a$$

$$= \frac{w_o''}{1 - \alpha}\, \sin\frac{\pi x}{L}$$

The delfection amplitude seems to be a product of two terms, the first term being the initial amplitude without axial load and the second term being an amplification factor which depends upon the value of $\alpha$.

It is to be expected that $w_o'''$ is maximum at the critical state of stress, but since only the initial irregularities, whose wave lengths are the same as the critical wave length of wrinkling, are considered in the present analysis, the most critical condition is produced when $\alpha$ becomes $(P/P_{cr})$, where $P_{cr}$ is the wrinkling load [7, 8, 22]. Thus, Equ. (5.18) can be written as:

$$w_o''' = \frac{w_o''}{\dfrac{\sigma_{cr}}{\sigma} - 1} \qquad (5.19)$$

As an increase in the applied compressive load will give rise to additional core deformations, loading to critical situation where the ultimate strength of the core

material is reached before wrinkling can occur, and thus results in an internal core failure. An analysis to determine the failing stress is given next.

Consider a panel subjected to compressive load. The maximum normal and shear strains in the core are:

$$\varepsilon_z = \left(\frac{\partial w_a}{\partial z}\right)_{\substack{z=0 \\ x=L/2}} = K\, w_o'''$$

$$\gamma_{xz} = \left(\frac{\partial w_a}{\partial x}\right)_{\substack{z=0 \\ x=0}} = \frac{\pi}{L}\, w_o'''$$

From Hooke's law, assuming that the stress-strain relations for the core material is linear up to failure point:

$$w_o''' = \frac{T}{E_c\, K}$$

or

$$w_o''' = \frac{L\, S}{\pi\, G_c}$$

in which
T = the core compressive or tensile strength;
S = the core shear strength.

By substituting these expressions into Equ. (5.19), it is found that:

$$\sigma_{fn} = \frac{\sigma_{cr}}{1 + \dfrac{E_c \, K \, w''_0}{T}}$$

$$\sigma_{fs} = \frac{\sigma_{cr}}{1 + \dfrac{\pi \, G_c \, w''_0}{L \, S}}$$

(5.20)

in which
$\sigma_{fn}$ = the failing stress of the core material based on its strength in compression or tension;
$\sigma_{fs}$ = same meaning as $\sigma_{fn}$ but for shear stress.

By means of Equ. (5.20), for any given initial imperfection and core strength, the failing stress can be determined, or conversely, for any given initial imperfection, the core strength required to sustain a specific axial load can be computed.

It should be noted that, since the core material is stressed beyond its proportional limit, the so-called plasticity factor should be used in calculating the wrinkling stress in Equs. (5.20). Thus

$$\sigma_{cr} = K_w(\eta E_f E_c G_c)^{1/3}$$
(5.21)

in which
$\eta$ = the plasticity factor for facing material, when subjected to edge compression. A formula for this factor can be obtained from References [7, 15, 18, 22].

## 5.4   CRIMPING LOAD OF COMPOSITE PANELS

Consider a panel (Fig. 5.6) under axial load P. The total deflection w is composed of two superimposed components: $w_b$ due to flexure and $w_s$ due to shear, as shown in Fig. 5.6(a). The slope of the deflected surface can be obtained from:

$$\frac{dw}{dx} = \frac{dw_b}{dx} + \frac{dw_s}{dx}$$

$$= \left( \int_o^x - \frac{M}{D} dx \right) + \frac{Q}{S} \tag{5.22}$$

in which

M   =   the applied bending moment on a section at a distance x from the origin (Fig. 5.6(b))

=   Pw;

Q   =   the shear force at the same section

=   P(dw/dx);

S   =   the shear stiffness given in chapter II.

Differentiating Equ. (5.22) once with respect to x and substituting the expressions for M and Q resulted in:

$$\frac{d^2w}{dx^2} = \frac{P/D}{\frac{P}{S} - 1} w \tag{5.23}$$

A solution for this equation must satisfy the following boundary conditions:

w   =   0

$d^2w_b/dx^2$   =   0         at x = 0 and x = a

and is taken in the form:

$$w = \sum_{m=1}^{\infty} C_1 \sin \alpha_m x + C_2 \cos \alpha_m x + C_3$$

in which

$$\alpha_m = \frac{m\pi}{a} \qquad m = 1,2,\ldots$$

$C_1, C_2, C_3$ = coefficients to be determined by satisfying the boundary conditions.

By satisfying the boundary conditions, it is found that:

$$w = \sum_{m=1}^{\infty} C_1 \sin \alpha_m x \qquad\qquad (5.24)$$

Substituting this solution into Equ. (5.23) and simplifying the terms gives:

$$P = \frac{m^2 P_E}{1 + m^2 \dfrac{P_E}{S}}$$

in which
$P_E$ = Euler load
    = $\pi^2 D/a^2$;
$D$ = the flexural rigidity of the strut;
$a$ = the length of the strut (Fig. 5.6(a)).

When $m \to \infty$ this formula becomes

$$P = S \qquad\qquad (5.25)$$

which is conventionally known as the crimping load.

## 5.5 REFERENCES

[1] Allen, H.G., Analysis and Design of Structural Sandwich Panels, First edition, Pergamon Press Ltd., 1969.

[2]    Chong, K.P., and Hartsock, J.A., "Flexural Wrinkling in Foam-Filled Sandwich Panels", J. Mech. Div., ASCE, Vol. 100, No. EM1, February, 1974.

[3]    Column Research Committee of Japan, Handbook of Structural Stability, Corona Publishing Company, Ltd., 1971.

[4]    Goodier, J.N., and Hsu, C.S., "Nonsinusoidal Buckling Modes of Sandwich Plates", J. Aero. Sci., Vol. 21, No. 8, August, 1954.

[5]    Gough, Elam and Debruyne, "The Instability of Thin Sheet by Supporting Medium", Journal of the Royal Aeronautical Society, Vol. XLIV, No. 349, quoted for [90], January, 1940.

[6]    The Computer Library at Concordia University, Montreal, Canada.

[7]    Harris, B.J., "Face-Wrinkling Mode of Buckling of Sandwich Panels", J. Eng. Mech. Div., ASCE, Vol. 91, No. EM3, June, 1965.

[8]    Harris, B.J., and Nordby, G.M., "Local Failure of Plastic-Foam Core Sandwich Panels", J. Struct. Div., ASCE, Vol. 95, No. ST4, April, 1969.

[9]    Hoff, N.J., and Mautner, S.E., "The Buckling of Sandwich-Type Panels", Journal of the Aeornautical Sciences, Vol. 12, No. 3, July, 1945.

[10]   Kuenzi, E.W., Norris, C.B., and Jenkinson, P.M., "Buckling Coefficient for Simply Supported and Clamped Flat, Rectangular Sandwich Panels Under Edgewise Compression", U.S. Forest Products Laboratory, Research Note FPL 070, December, 1964.

[11]   Lin, T.H., and Yokota, L.T., "Deflections and Bending Moments of Rectangular Sandwich Panels with Clamped Edges under Combined Biaxial Compressions and Pressure", J. American Inst. of Aeronautics and Astronautics, Vol. 3, No. 6, June, 1965.

[12]   Marsh, C., Theory of Elastic Stability - Lecture Notes, Concordia University, 1978.

[13]   McGlenn, J., and Hartz, B., "Finite Element Analysis of Plywood Plates", J. Struct. Div., ASCE, Vol. 94, No. ST2, February, 1968.

[14]   Plantema, F.J., Sandwich Construction - The Bending and Buckling of Sandwich Beams, Plates and Shells, John Wiley and Sons, Inc., 1966.

[15]   Sullins, R.T., Smith, G.W., and Spier, E.E., Manual for Structural Stability Analysis

of Sandwich Plates and Shells, NASA CR-1457, December, 1969.

[16] Szilard, R., Theory and Analysis of Plates - Classical and Numerical Methods, Prentice-Hall, Inc., 1974.

[17] Timoshenko, S., and Krieger, S., Theory of Plates and Shells, Second edition, McGraw-Hill Book Company, 1959.

[18] Timoshenko, S.P., and Gere, J.M., Theory of Elastic Stability, Second edition, McGraw-Hill Book Company, Inc., 1961.

[19] Vanderbilt, M.D., Goodman, J.R., and Criswell, M.E., "Service and Overload Behaviour of Wood Joist Floor Systems", J. Struct. Div., ASCE, Vol. 100, No. ST1, January, 1974.

[20] Williams, D., "Sandwich Construction - A Practical Approach for the Use of Designers", R.A.E. Report No. Structure 2, quoted from [90], April, 1947.

[21] Yusuff, S., "Theory of Wrinkling in Sandwich Construction", J. of the Royal Aeronautical Society, Vol. 59, January, 1955.

[22] Yusuff, S., "Face Wrinkling and Core Strength in Sandwich Construction", J. of the Royal Aeronautical Society, Vol. 64, March, 1960.

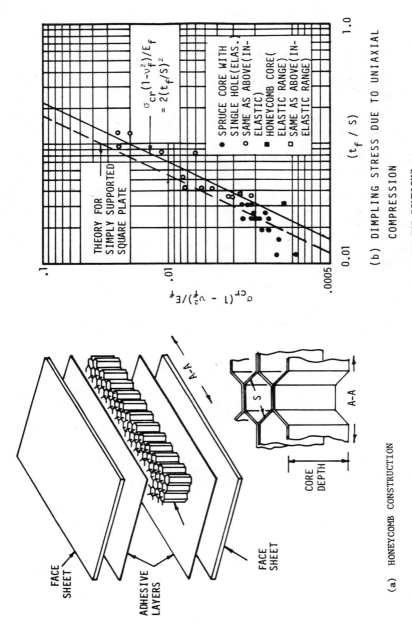

(b) DIMPLING STRESS DUE TO UNIAXIAL
COMPRESSION

(a) HONEYCOMB CONSTRUCTION

Fig. 5.1 HONEYCOMB COMPOSITE PLATE AND EXPERIMENTAL DATA FOR DIMPLING

168

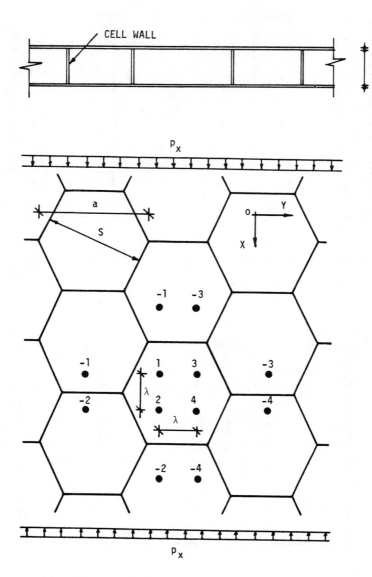

Fig. 5.2 FINITE DIFFERENCE MESH FOR HONEYCOMB PLATE

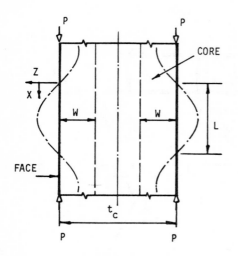 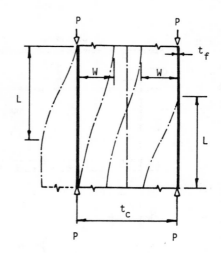

(a) SYMMETRIC WRINKLING          (b) ANTISYMMETRIC WRINKLING

Fig. 5.3 - FACE WRINKLING

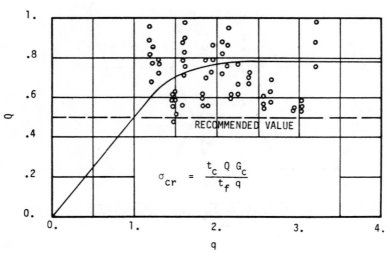

$$\sigma_{cr} = \frac{t_c \, Q \, G_c}{t_f \, q}$$

RECOMMENDED VALUE

Fig. 5.4 - RECOMMENDED VALUE FOR WRINKLING COEFFICIENT

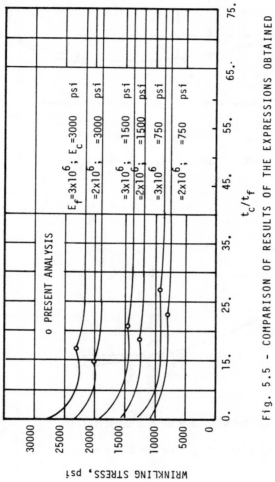

Fig. 5.5 - COMPARISON OF RESULTS OF THE EXPRESSIONS OBTAINED
TO DEFINE WRINKLING TRANSITION STATE WITH HOFF'
THEORY

171

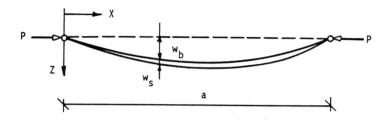

(a)    COMPOSITE PANEL UNDER COMPRESSIVE LOAD

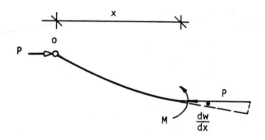

(b)    FREE BODY DIAGRAM OF A COMPOSITE PANEL

Fig. 5.6   CRIMPING LOAD OF COMPOSITE PANELS

# VI

# Hygrothermal Stresses in Composite Panels

## 6.1  INTRODUCTION

Structural systems are usually called upon to experience thermal stresses when they are subjected to a temperature change. There are three causes for the thermally induced stresses. First of all, when the temperature change in a homogeneous pate, for instance, is not uniform, the different elements of its body tend to deform differently, but since the restriction of displacement continuity will prevent such free deformations, it follows that various elements exert upon each other a restraining action which produces a system of strains to cancel out a part of the free deformations at every point so as to ensure continuity of the displacement This system of strains must be accompanied by a corresponding system of stresses which are known as thermal stresses. Second, when the temperature change in a homogeneous plate is uniform, but the free displacements are limited by certain boundary conditions, thermal stresses will be imposed. Finally, when a plate is heterogeneous or made of dissimilar materials, then it may be unable to deform freely in a manner compatible with the temperature distribution through it, and hence thermal stresses will be exhibited, although the plate may be free to move at its boundaries [5, 13, 15, 32]. Based on this discussion, composite components will undergo thermal stresses due to uniform and gradient temperature changes, whether the boundaries are free or not.

In this chapter a solution for thermally induced stresses in composite plates subjected to a uniform temperature change is developed using an elasticity approach. The finite element method is used for comparison. The 24 x 24 stiffness matrix of an orthotropic rectangular prism finite element is derived and used in the finite element computer program developed based on the direct stiffness method. To facilitate the use of this solution, simple formulas are derived to calculate the maximum normal stress in facings and interlayer shear in panels subjected to uniform temperature change. Based on engineering theory, simple formulas are presented to calculate the maximum normal stresses in facings of composite panels subjected to a moisture content change. The results of existing theories in the literature for thermal stresses in plates subjected to thermal gradient are reformulated in this chapter and presented in simple forms for practical applications.

With regard to the theories available in this area, several solutions have been found for thermal stresses in plates subjected to thermal gradient. Bijlaard [4] and Jerzy [20] used the theory of elasticity to solve plates subjected to thermal gradient and with the following boundary conditions:  plate with clamped edges, plate with simply supported edges, plate with two opposite edges simply supported and remaining edges free, plate with two opposite edges simply supported and remaining edges free but with edge stiffeners.  According to the results by Bijlaard and Jerzy, the states of deformation and stresses for a plate with simply supported or clamped edges and subjected to thermal gradient are the same as for homogeneous plates.  By using the principle of minimum total potential energy, Ebcioglu [7, 11] solved simply supported plates subjected to transverse load, edge compression and thermal gradient.

The problem of thermal stresses has also been investigated experimentally and by using the finite element method.  Panels with cold-formed steel facings and subjected to a thermal gradient across the thickness were used in a series of experiments conducted by Chong [8] to measure the deflection and stresses induced.  The results were compared with analytical expressions derived based on a fourth order governing differential equation established for the deflection.  Marsh [23, 24] investigated experimentally and by using an elasticity approach the thermal stresses in strips subjected to uniform temperature change and thermal gradient.  By using the finite element, the transverse displacement of an axisymmetric heated cylinder was determined by Monforton [28].  On the other hand, the hybrid finite element is used by Hoa [16] to investigate the thermal stresses in homogeneous plates.

## 6.2    THERMAL STRESSES IN COMPOSITE PLATES SUBJECTED TO A UNIFORM TEMPERATURE CHANGE

### 6.2.1    ASSUMPTIONS

Consider a composite plate composed of two facings each of thickness $t_f$, a core of thickness $t_c$, and subjected to a uniform temperature change as shown in Fig. 6.1.  As the need to consider the thermal deformations of different materials necessitates the consideration of compatibility of deformations and stresses at the interfaces, the problem is irreducibly three dimensional.  Several assumptions mentioned in Chapter IV can still be retained, and new ones are introduced.  These are:
(1) Inplane shear stresses in the facings can be ignored.
(2) The temperature distribution is constant in the plane of each face and across the plate thickness.
(3) The plate is free to expand along its edges.
(4) The temperature level does not affect the material properties.

Assumption (1) effectively implies the existence of a simplified two dimensional stress state in the facings.  Assumption (3) represents an idelaized form of behavior of free edges composite plate.  This assumption was considered by Marsh in his analysis [23, 24].  Considering constant temperature distribution in the plane of the plate is commonly used in practice [4, 20].

### 6.2.2 GENERAL

When a plate is subject to a temperature change, the different laminae tend to deform differently. Since the restriction of displacement continuity will prevent such free deformations, it follows that various laminae exert upon each other a restraining action which induces a system of strains accompanied by a system of thermal stresses.

Consider now the facings, which are assumed to be in a plane stress condition with the non-zero stress components $\sigma_{fx}$, $\sigma_{fy}$, and $\tau_{xy}$. The differential equations of equilibrium for the facings are:

$$\frac{\partial \sigma_{fx}^j}{\partial x} + \frac{\partial \tau_{yx}^j}{\partial y} \; \bar{+} \; \frac{q_x}{t_f^j} \; = \; 0 \qquad\qquad (4.2)$$

$$\frac{\partial \sigma_{fy}^j}{\partial y} + \frac{\partial \tau_{xy}^j}{\partial x} \; \bar{+} \; \frac{q_y}{t_f^j} \; = \; 0 \qquad\qquad (4.3)$$

in which

j    =  subscript denoting stresses in the face j, while $t_f^j$ is the thickness of this face;

f    =  subscript denoting face;

$q_x, q_y$ =  interlayer shear stresses.

The negative and positive signs correspond to j = 1 for the bottom face, and j = 2 for the upper one, respectively.

The state of stresses in the core must satisfy the following equilibrium equations:

$$\frac{\partial \sigma_{cx}}{\partial x} + \frac{\partial \tau_{czx}}{\partial z} = 0$$

$$\qquad\qquad (4.4)$$

$$\frac{\partial \sigma_{cy}}{\partial y} + \frac{\partial \tau_{czy}}{\partial z} = 0$$

in which

$\sigma_{ci}$ = normal stress in the core in the i- direction;

$\tau_{czi}$ = shear stress in the core;

i = subscript denoting x or y.

The equilibrium equations in terms of displacement components are:

$$E_{cx} \frac{\partial^2 u_c}{\partial x^2} + G_{cx} \frac{\partial^2 u_c}{\partial z^2} + G_{cx} \frac{\partial^2 w}{\partial z \partial x} = 0$$

$$\tag{6.1}$$

$$E_{cy} \frac{\partial^2 v_c}{\partial y^2} + G_{cy} \frac{\partial^2 v_c}{\partial z^2} + G_{cy} \frac{\partial^2 w}{\partial z \partial y} = 0$$

in which

$u_c, v_c$ = displacement components in the core along x- and y- directions, respectively;

E, G = the elastic and shear modulii of the core material, respectively;

c = subscript denoting core;

w = the lateral deflection.

At the interfaces between the core and the faces, the stresses and deformations must be compatible. The compatibility equations in terms of stresses are:

$$q_x = G_{cx} \left( \frac{\partial u_c}{\partial z} + \frac{\partial w}{\partial x} \right)_{z=t_c/2}$$

$$\tag{4.8}$$

$$q_y = G_{cy} \left( \frac{\partial v_c}{\partial z} + \frac{\partial w}{\partial y} \right)_{z=t_c/2}$$

In terms of displacement components, the compatibility equations are written as [18, 23, 24]:

$$u_f - (u_c)_{z=t_c/2} = \Delta x$$

$$\tag{6.2}$$

$$v_f - (v_c)_{z=t_c/2} = \Delta y$$

in which

$$\Delta \quad = \quad (\alpha_c - \alpha_f)T$$

where $\alpha_c$ and $\alpha_f$ are the thermal expansion coefficients of the core and faces materials, respectively. T is the temperature change at the plate surfaces;

$u_f$, $v_f$ = displacement components in facings along x- and y- directions, respectively.

Solutions to the problem must also satisfy the boundary conditions. With regard to the plate in Fig. 6.1, the relevant boundary conditions are:

$$
\begin{array}{llll}
\text{at } x = 0 & u_f = u_c = 0 & \text{and} & \tau_{xz} = 0 \\
x = \pm a & \sigma_{fx} = 0 & \text{and} & \sigma_{cx} = 0 \\
\text{at } y = 0 & v_f = v_c = 0 & \text{and} & \tau_{yz} = 0 \\
y = \pm b & \sigma_{fy} = 0 & \text{and} & \sigma_{cy} = 0 \\
\text{at } z = 0 & & \tau_{zx} = & \tau_{zy} = 0
\end{array}
\tag{6.3}
$$

These equations imply that no normal stress exists in the core or facings along the boundaries. Due to the symmetry in the panel geometry and temperature distributions, the transverse shear stress $\tau_{xz}$, $\tau_{yz}$ and the displacement components $u_c$, $v_c$, $u_f$, and $v_f$ vanish at the transverse central planes. Because the two facings are identical, no horizontal shear stresses exist on the middle plane.

### 6.2.3  THERMAL STRESSES

Consider the plate in Fig. 6.1 subjected to a uniform temperature change. The equilibrium equations (6.1) are:

$$
E_{cx} \frac{\partial^2 u_c}{\partial x^2} + G_{cx} \frac{\partial^2 u_c}{\partial z^2} = 0
\tag{4.4}
$$

$$
E_{cy} \frac{\partial^2 v_c}{\partial y^2} + G_{cy} \frac{\partial^2 v_c}{\partial z^2} = 0
\tag{4.5}
$$

The solution satsifying the boundary conditions of the core layer in Equ. (6.3) is found as [18, 23, 24]:

$$u_c = \sum_{m=1,3,\ldots}^{\infty} C_m \, \sin \alpha_m x \, \cosh \mu_x \, \alpha_m z$$

$$\text{(6.4)}$$

$$v_c = \sum_{n=1,3,\ldots}^{\infty} D_n \, \sin \beta_n y \, \cosh \mu_y \, \beta_n z$$

in which

| | | |
|---|---|---|
| $\alpha_m$ | $=$ | $m\pi/2a$ |
| $\beta_n$ | $=$ | $n\pi/2b$ |
| $\mu_x$ | $=$ | $(E_{cx}/G_{cx})^{1/2}$ |
| $\mu_y$ | $=$ | $(E_{cy}/G_{cy})^{1/2}$ |
| $C_m, D_n$ | $=$ | unknown functions of y and x, respectively |
| where | | |
| 2a, 2b | $=$ | plate dimensions along x- and y- axes, respectively. |

The transverse shear stresses in the core are related to $u_c$ and $v_c$ by the following equations:

$$\tau_{xz} = G_{cx} \frac{\partial u_c}{\partial z}$$

$$\tau_{yz} = G_{cy} \frac{\partial v_c}{\partial z}$$

Thus, substituting Equ. (6.4) in these relations gives:

$$\tau_{xz} = G_{cx} \sum_{m=1,3,\ldots}^{\infty} \mu_x \alpha_m C_m \sin \alpha_m x \sinh \mu_x \alpha_m z \qquad (6.5)$$

$$\tau_{yz} = G_{cy} \sum_{n=1,3,\ldots}^{\infty} \mu_y \beta_n D_n \sin \beta_n y \sinh \mu_y \beta_n z \qquad (6.6)$$

from which, the interlayer shear stresses are:

$$q_x = G_{cx} \sum_{m=1,3,\ldots}^{\infty} \mu_x \alpha_m C_m \sin \alpha_m x \sinh \frac{1}{2} \mu_x \alpha_m t_c$$

$$q_y = G_{cy} \sum_{n=1,3,\ldots}^{\infty} \mu_y \beta_n D_n \sin \beta_n y \sinh \frac{1}{2} \mu_y \beta_n t_c$$

These equations together with Equs. (4.2) and (4.3); noting that $\tau_{xy} = \tau_{yx} = 0$, $\sigma_{fx} = 0$ at x = ±a and $\sigma_{fy} = 0$ at y = ±b; give:

$$\sigma_{fx} = -\mu_x \frac{G_{cx}}{t_f} \sum_{m=1,3,\ldots}^{\infty} C_m \cos \alpha_m x \sinh \frac{1}{2} \mu_x \alpha_m t_c \qquad (6.7)$$

$$\sigma_{fy} = -\mu_y \frac{G_{cy}}{t_f} \sum_{n=1,3,\ldots}^{\infty} D_n \cos \beta_n y \sinh \frac{1}{2} \mu_y \beta_n t_c \qquad (6.8)$$

The displacement components in the facings can be obtained from Equs. (6.7) and (6.8) together with the two dimensional Hookes law as:

$$u_f = \sum_{m=1,3,\ldots}^{\infty} - C_m \frac{\Psi}{\alpha_m} \sinh \frac{1}{2} \mu_x \alpha_m t_c \sin \alpha_m x + \nu \Psi' \xi' x$$

$$v_f = \sum_{n=1,3,\ldots}^{\infty} - D_n \frac{\Psi'}{\beta_n} \sinh \frac{1}{2} \mu_y \beta_n t_c \sin \beta_n y + \nu \Psi \xi y$$

in which

$$\Psi = \frac{\mu_x G_{cx}}{E_f t_f}$$

$$\Psi' = \frac{\mu_y G_{cy}}{E_f t_f}$$

$$\xi' = \sum_{n=1,3,\ldots}^{\infty} D_n \cos \beta_n y \sinh \frac{1}{2} \mu_y \beta_n t_c$$

$$\xi = \sum_{m=1,3,\ldots}^{\infty} C_m \cos \alpha_m x \sinh \frac{1}{2} \mu_x \alpha_m t_c$$

By proper substitution of $u_c$, $v_c$, $u_f$, and $v_f$ in the compatibility equations (6.2), it is found that:

$$\sum_{m=1,3,\ldots}^{\infty} C_m \left[ (\frac{\Psi}{\alpha_m} \sinh \frac{1}{2} \mu_x \alpha_m t_c) + \cosh \frac{1}{2} \mu_x \alpha_m t_c \right] \sin \alpha_m x =$$

$$x (\nu \Psi' \xi' - \Delta)$$

$$\sum_{n=1,3,\ldots}^{\infty} D_n \left[ (\frac{\Psi'}{\beta_n} \sinh \frac{1}{2} \mu_y \beta_n t_c) + \cosh \frac{1}{2} \mu_y \beta_n t_c \right] \sin \beta_n y =$$

$$y (\nu \Psi \xi - \Delta)$$

Expanding x and y in the right hand side of these equations by single Fourier series in the sine terms, and equating the coefficients in both sides, expressions for $C_m$ and $D_n$ are determined as:

$$C_m = \frac{2 \, (-1)^{(m-1)/2}}{a \, \alpha_m^2} \cdot \frac{(\nu \, \Psi' \, \xi' - \Delta)}{\frac{\Psi}{\alpha_m} \sinh \frac{1}{2} \mu_x \alpha_m t_c + \cosh \frac{1}{2} \mu_{x.} \alpha_m t_c}$$

$$D_n = \frac{2 \, (-1)^{(n-1)/2}}{b \, \beta_n^2} \cdot \frac{(\nu \, \Psi \, \xi - \Delta)}{\frac{\Psi'}{\beta_n} \sinh \frac{1}{2} \mu_y \beta_n t_c + \cosh \frac{1}{2} \mu_y \beta_n t_c}$$
(6.9)

Equivalent expressions can be obtained for $\xi$ and $\xi'$ by multiplying the first equation in (6.9) by $\cos \alpha_m x \sinh (1/2)\mu_x \alpha_m t_c$, hence

$$\xi = \theta \, (\nu \, \Psi' \, \xi' - \Delta) \tag{6.10}$$

in which

$$\theta = \sum_{m=1,3,\ldots}^{\infty} \frac{2 \, (-1)^{(m-1)/2}}{a \, \alpha_m^2} \cdot \frac{\cos \alpha_m x \, \sinh \frac{1}{2} \mu_x \alpha_m t_c}{\frac{\Psi}{\alpha_m} \sinh \frac{1}{2} \mu_x \alpha_m t_c + \cosh \frac{1}{2} \mu_x \alpha_m t_c}$$

in a similar manner, from the second equation in (6.9), it is found that:

$$\xi' = \theta' \, (\nu \, \Psi \, \xi - \Delta) \tag{6.11}$$

in which

$$\theta' = \sum_{n=1,3,\ldots}^{\infty} \frac{2 \, (-1)^{(n-1)/2}}{b \, \beta_n^2} \cdot \frac{\cos \beta_n y \, \sinh \frac{1}{2} \mu_y \beta_n t_c}{\frac{\Psi'}{\beta_n} \sinh \frac{1}{2} \mu_y \beta_n t_c + \cosh \frac{1}{2} \mu_y \beta_n t_c}$$

Solving for $\xi$ and $\xi'$ from Equs. (6.10) and (6.11) yields:

$$\xi = \Delta\theta \; \frac{1 + \nu\,\Psi'\,\theta'}{\nu^2\,\Psi\,\Psi'\,\theta\,\theta' - 1}$$

$$\xi' = \Delta\theta' \; \frac{1 + \nu\,\Psi\,\theta}{\nu^2\,\Psi\,\Psi'\,\theta\,\theta' - 1}$$

The complexity of the above analysis justifies now examining the simplified case of strips, that is when $2b \to \infty$. In this case $\beta_n \to 0$. Thus:

$$v_c = \tau_{yz} = 0$$
$$\xi' = \theta' = 0$$

and the first equation in (6.9) becomes

$$C_m = -\frac{2\,(-1)^{(m-1)/2}}{a\,\alpha_m^2} \; \frac{\Delta}{\dfrac{\Psi}{\alpha_m}\,\sinh\frac{1}{2}\mu_x\,\alpha_m\,t_c + \cosh\frac{1}{2}\mu_x\,\alpha_m\,t_c}$$

which is in exact agreement with the result obtained by Marsh [23, 24].

## 6.3  THERMAL STRESSES BY THE FINITE ELEMENT METHOD

In the analysis of structures, the finite element method has wide acceptance by engineers. However, the basic concept of this method is that a continuum can be modeled analytically through its subdivision into elements, the behavior of each is described with the aid of either a stress or a displacement function. The method has a theoretical basis within the framework of the classical theory and is related to the Ritz's method applied in solving structural problems.

In the conventional Ritz procedure, the displacement field in the entire continuum is described by one set of functions, nevertheless, in the finite element methods, the displacement field for each element is defined individually in terms of the displacements at its nodes or points of connections, and hence the continuity is assured within the separate elements whereas the whole domain can be regarded as piece-wise and continuous. However, the continuity preserved at the nodes by definition, while a careful choice of the displacement field of the elements may assure the continuity across element boundaries as well.

Research in the field of the finite element idealization has produced a large number of basic elements to fit a particular purpose or to be applied in solving a specific problem [1, 2, 3, 6, 9, 10, 12, 14, 16, 17, 19, 21, 22, 25, 26, 27, 28, 29, 30, 31, 32, 33, 34, 35]. As the present work is concerned with the thermal stresses in a composite plate subjected to a temperature change, the search will be directed towards the orthotropic three dimensional finite elements. In this category, the rectangular prism with a linear variation of the displacements along its edges seems to be reasonably acceptable, since it endures the displacement compatibility at the element interfaces and was proved to produce better results than others [27].

## 6.3.1  FINITE ELEMENT ANALYSIS

To solve a structural problem by applying the finite element method, three conditions are to be satisfied: geometrical, mechanical, and static conditions. First, once the displacement field is described within an element in terms of its nodal displacements, the strain components can be determined readily by the proper differentiation of the assumed displacement function, that is:

$$\{\varepsilon\} \quad = \quad [B]\{d\} \tag{6.12}$$

in which

  $\{\varepsilon\}$  =  the strain components vector;
  $\{d\}$  =  the nodal displacements vector;
  $[B]$  =  the strain-displacement matrix.

and this equation represents the geometric relation of compatibility.

Second, the constitutive equation which relates stresses to strains is used to express the second condition, hence

$$\{\sigma\} \quad = \quad [E]\{\varepsilon\} \tag{6.13}$$

in which

  $\{\sigma\}$  =  the stress components vector;
  $[E]$  =  the elasticity matrix.

Finally, the third condition is satisfied now through the use of the theorem of stationary total potential energy which states: "Among the numerous sets of compatible displacements, the one which also satisfies equilibrium will extremize the total potential energy of the elastic system".

Thus

$$\frac{\partial \Pi}{\partial \{D\}} = 0 \qquad\qquad (6.14)$$

in which

$\pi$ = the total potential energy function of the structural system,

$$= \sum_{i=1}^{n} \Pi_i$$

where, $\pi_i$ is the total potential energy of the $i^{th}$ finite element

$\{D\}$ = the structural nodal degrees of freedom.

By following the conventional procedure of the assumed displacement finite element method [10, 12, 17, 27, 35], it is concluded that:

$$\pi_i = U_i + V_i$$

in which

$U_i$ = the element strain energy,

= $(1/2) \int_{vol.} \{d\}_i^T [B]_i^T [E]_i [B]_i \{d\}_i dv$

where dv is the element volume

$V_i$ = the potential of the nodal loads $\{r\}_i$

= $-\{r\}_i^T \{d\}_i$

consequently, it follows from Equ. (6.14) that:

$$\sum_{i=1}^{n} [K]_i \{d\}_i = \sum_{i=1}^{n} \{r\}_i$$

or in an abbreviated form

$[K]\{D\} = \{Q\}$ $\qquad\qquad$ (6.15)

in which

$[K]_i$ = the element stiffness matrix

= $\int_{vol.} [B]_i^T [E]_i [B]_i \, dv$            (6.16)

$[K]$ = the structural stiffness matrix

= $\sum\limits_{i=1}^{n} [K]_i$

$\{Q\}$ = the structural nodal forces

= $\sum\limits_{i=1}^{n} \{r\}_i$

The preceding formulation is the general description step by step of the finite element method. For the particular case of the rectangular prism in Fig. 6.2, the details are given next. The displacement field within the element is defined by [35]

$$\{f\} = \begin{array}{c} u \\ v \\ w \end{array}$$

in which

u, v, w = the displacement components in X-, Y-, and Z- directions, respectively;

$\{f\}$ = the displacement within the element.

which can be written in terms of the nodal degrees of freedom as

$$\{f\} = [IN_1 \; IN_2 \; IN_3 \; IN_4 \; IN_5 \; IN_6 \; IN_7 \; IN_8] \{d\} \qquad (6.17)$$

in which

I = identity matrix of order three

$\{d\}$ = the element nodal degrees of freedom

= $\{u_1 \; v_1 \; w_1 \dots\dots u_8 \; v_8 \; w_8\}^T$

where 1 to 8 are a node's number of the element (Fig. 6.2)

$N_{,,}$ = shape function at node $,,$ ($,,$ = 1 to 8)

= $\frac{1}{8} (1 + \xi \, \xi_\ell)(1 + \eta \, \eta_\ell)(1 + \varsigma \, \varsigma_\ell)$

where

$$\xi = x/a, \qquad \eta = y/b, \qquad \zeta = z/c$$

$\xi_n$, $\eta_n$, $\zeta_n$ take the values $\pm 1$ depending on the node position relative to the coordinate axis as shown in Fig. 6.2, and 2a, 2b, and 2c = the element dimensions parallel to x-, y-, and z- axes, respectively.

By choosing the displacement function, the strain components can be determined according to

$$\{\varepsilon\} = \begin{Bmatrix} \varepsilon_x \\ \varepsilon_y \\ \varepsilon_z \\ \gamma_{xy} \\ \gamma_{yz} \\ \gamma_{zx} \end{Bmatrix} = \begin{Bmatrix} \dfrac{\partial u}{\partial x} \\[6pt] \dfrac{\partial v}{\partial y} \\[6pt] \dfrac{\partial w}{\partial z} \\[6pt] \dfrac{\partial u}{\partial y} + \dfrac{\partial v}{\partial x} \\[6pt] \dfrac{\partial v}{\partial z} + \dfrac{\partial w}{\partial y} \\[6pt] \dfrac{\partial w}{\partial x} + \dfrac{\partial u}{\partial z} \end{Bmatrix} \qquad (6.18)$$

in which
$\varepsilon, \gamma$ = normal and shear strains, respectively.

and consequently the strain-displacement matrix is obtained from Equs. (6.17) and (6.18) as:

$$[B] = [B_1 \ B_2 \ B_3 \ B_4 \ B_5 \ B_6 \ B_7 \ B_8] \qquad (6.19)$$

in which

$B_h$ = a six by three matrix. The explicit forms of $B_1$ to $B_8$ are given in Appendix E (h = 1 to 8).

The general Hookes law for orthotropic material is written in matrix form as:

$$\{\varepsilon\} = [C]\{\sigma\}$$

in which

[C] = a square matrix of order six which includes the elastic properties of the finite element material

$\{\sigma\}$ = stress components corresponding to the strain $\{\varepsilon\}$.

The elasticity matrix [E] in Equ. (6.13), thus, is given by

$$[E] = [C]^{-1}$$

$$= \begin{array}{cc} E_{11} & 0 \\ 0 & E_{22} \end{array} \tag{6.20}$$

in which

$$[E_{11}] = \frac{1}{e} \begin{bmatrix} \dfrac{1}{E_y E_z} - \dfrac{\nu_{zy}}{E_z}\dfrac{\nu_{yz}}{E_y} & \dfrac{1}{E_z}\dfrac{\nu_{yx}}{E_y} + \dfrac{\nu_{zx}}{E_z}\dfrac{\nu_{yz}}{E_y} & \dfrac{1}{E_y}\dfrac{\nu_{zx}}{E_z} + \dfrac{\nu_{yx}}{E_y}\dfrac{\nu_{zy}}{E_z} \\[3mm] \dfrac{1}{E_z}\dfrac{\nu_{xy}}{E_x} + \dfrac{\nu_{xz}}{E_x}\dfrac{\nu_{zy}}{E_z} & \dfrac{1}{E_x E_z} - \dfrac{\nu_{zx}}{E_z}\dfrac{\nu_{xz}}{E_x} & \dfrac{1}{E_x}\dfrac{\nu_{zy}}{E_z} + \dfrac{\nu_{zx}}{E_z}\dfrac{\nu_{xy}}{E_x} \\[3mm] \dfrac{1}{E_y}\dfrac{\nu_{xz}}{E_x} + \dfrac{\nu_{xy}}{E_x}\dfrac{\nu_{yz}}{E_y} & \dfrac{1}{E_x}\dfrac{\nu_{yz}}{E_y} + \dfrac{\nu_{yx}}{E_y}\dfrac{\nu_{yz}}{E_x} & \dfrac{1}{E_x E_y} - \dfrac{\nu_{yx}}{E_y}\dfrac{\nu_{xy}}{E_x} \end{bmatrix}$$

$$[E_{22}] = \begin{array}{ccc} G_{xy} & 0 & 0 \\ 0 & G_{yz} & 0 \\ 0 & 0 & G_{zx} \end{array}$$

where

$$e = \frac{1}{E_x E_y E_z} - \frac{1}{E_x} \frac{\nu_{zy}}{E_z} \frac{\nu_{yz}}{E_y} - \frac{1}{E_z} \frac{\nu_{xy}}{E_x} \frac{\nu_{yx}}{E_y}$$

$$- \frac{\nu_{xy}}{E_x} \frac{\nu_{zx}}{E_z} \frac{\nu_{yz}}{E_y} - \frac{\nu_{xz}}{E_x} \frac{\nu_{yx}}{E_y} \frac{\nu_{zy}}{E_z}$$

$$- \frac{1}{E_y} \frac{\nu_{xz}}{E_x} \frac{\nu_{zx}}{E_z}$$

| | | |
|---|---|---|
| $E_x, E_y, E_z$ | = | elastic moduli along X-, Y-, and Z- axes, respectively |
| $\nu_{ab}$ | = | Poisson's ratio, which characterizes the decrease in b-direction during tension applied in a- direction, where a and b can take x, y, or z. |
| $G_{ab}$ | = | the shear modulus which characterizes the change in the angle between a and b axes. |

Substituting Equs. (6.19) and (6.20) in Equ. (6.16) and performing the triple matrix product $[B]^T [E] [B]$ resulted in a 24 x 24 matrix populated by expressions which are function of $\xi$, $\eta$, and $\zeta$. Thus, integration of these expressions will lead to subsequent evaluation of the $K_{cd}$ terms of the element stiffness matrix $[K]_i$, and after proper simplification it is found that

$$[K]_i = \begin{matrix} K_{11} & & & \\ K_{21} & K_{22} & \text{Symmetric} & \\ K_{31} & K_{32} & K_{33} & \\ K_{41} & K_{42} & K_{43} & K_{44} \end{matrix} \qquad (6.21)$$

in which
$K_{cd}$ = a square matrix of order six.

By obtaining the element stiffness matrix, the structural stiffness matrix can be generated by simple summation of the individual element stiffness coefficients for common degrees of freedom at any node. This approach is often referred to as the direct stiffness method, and by which a set of simultaneous linear algebraic equations will be obtained, relating loads to displacement through the structural matrix as presented in Equ. (6.15). The temperature effect in this analysis is included in the load vector $\{r\}_i$ [10]:

$$\{r\}_i \quad = \quad \int_{vol.} [B]_i^T [E] \{\varepsilon_o\} \, dv \qquad\qquad (6.22)$$

in which

$\{\varepsilon_o\}$ = initial strain vector

= $\{\alpha T \quad \alpha T \quad \alpha T \quad o \quad o \quad o\}^T$
where T is the average change of temperature at the element nodes, i.e.

$$T = \frac{1}{8} \sum_{\ell=1}^{8} T_\ell$$

($\ell$ denoting a node number)

It should be noted that, since the element stiffness matrix is derived with respect to the selected natural coordinates which are the same as the global coordinates, no transformation is needed. The next step is to account for the appropriate boundary conditions in the structural stiffness matrix. In the present problem these are the symmetry conditions at the central planes $x = 0$, $y = 0$, and $z = 0$ of the plate in Fig. 6.1. The modified equation (6.15) can now be solved for the unknown displacements. When the nodal displacements have been determined, the element stresses follow from Hookes law:

$$\{\sigma\} \quad = \quad [E] \, (\{\varepsilon\} - \{\varepsilon_o\})$$

$$= \quad [E] \, ([B] \{d\} - \{\varepsilon_o\}) \qquad\qquad (6.23)$$

## 6.4 PRACTICAL FORMULAS FOR THERMAL STRESSES IN COMPOSITE PLATES

The analytic solutions developed for thermal stresses in composite panels with different boundary conditions [4, 7, 8, 11, 18, 20, 23, 24] are characterized by complex mathematical expressions which could limit their practical application. Consequently, to facilitate the use of these solutions, simple expressions are derived for the maximum deflection and stresses in plates subjected to temperature changes.

### 6.4.1   SIMPLY   SUPPORTED   PLATES SUBJECTED TO A THERMAL GRADIENT ±T

The analytic solution in this case was obtained by Bijlaard [4] and Jerzy [20]. As stated in their works the maximum thermal moments occur at the boundary of the plate and are given by

$$M = (M_x)_{max.} = (M_y)_{max.}$$

$$= \frac{\alpha \, T \, D \, (1-\nu^2)}{(t_c + t_f)} \tag{6.24}$$

in which

$\alpha$ = thermal expansion coefficient of the faces material;
T = temperature at the plate surfaces;
$\nu$ = Poisson's ratio of faces material;
D = flexural rigidity of the plate

$$= \frac{E_f \, t_f \, (t_c + t_f)^2}{2 \, (1-\nu^2)}$$

Substituting the expression for D into Equ. (6.24) gives

$$M = \frac{1}{2} \alpha \, T \, E_f \, t_f \, (t_c + t_f)$$

from which the maximum normal stress in the faces is determined from

$$(\sigma_f)_{max.} = \frac{M}{t_f \, (t_c + t_f)}$$

as

$$(\sigma_f)_{max.} = \frac{1}{2} \, \alpha \, T \, E_f \qquad\qquad (6.25)$$

The maximum deflection in this case was determined as

$$(w)_{max.} = \frac{4 \, \alpha \, T \, a^2 \, (1+\nu)}{\pi^3 \, (t_c + t_f)} \, K_w \qquad\qquad (6.26)$$

in which

$$K_w = \sum_{m=1,3,\ldots}^{\infty} \frac{(-1)^{(m-1)/2}}{m^3} \, (1 - \text{sech} \, \alpha_m)$$

where

$\alpha_m$ = $m\pi b/2a$

a = the plate length

b = the plate width

Numerical values for $K_w$ are obtained by the author for a range of (b/a) = .5 to 4. The results are presented in Fig. 6.3.

The maximum shear stress in this case occurs at the plate corners and has an infinite value. Bijlaard [4] used an approximate method to determine the maximum transverse shear force in the plate. He made use of the twisting moment distributions along the edges in his calculations.

## 6.4.2 CLAMPED EDGES PLATES SUBJECTED TO A THERMAL GRADIENT ± T

The analytic solution in this case was found by Bijlaard [4] and Jerzy [20]. The maximum thermal moments were determined as

$$M = (M_x)_{max.} = (M_y)_{max.}$$

$$= \frac{\alpha \, T \, D \, (1+\nu)}{(t_c + t_f)} \tag{6.27}$$

By following the same procedure adopted in the previous section, the maximum normal stress in the faces is obtained as

$$(\sigma_f)_{max.} = \frac{\alpha \, T \, E_f}{2 \, (1-\nu)} \tag{6.28}$$

A clamped plate will remain flat when it is exposed to a thermal gradient, thus

$$w(x, y) = 0$$

The shear forces are also vanished in this case.

### 6.4.3   PLATES SUBJECTED TO A UNIFORM TEMPERATURE CHANGE (Fig. 6.1)

Normal and interlayer shear stresses induced in plates subjected to uniform temperature change can be obtained from Equs. (6.7), (6.8), and (4.8) as:

$$\sigma_{fx} = (\sigma_{fx})_{\substack{x=0 \\ y=0}}$$

$$= - \mu_x \frac{G_{cx}}{t_f} \sum_{m=1,3,\ldots}^{\infty} C_m \sinh \frac{1}{2} \mu_x \alpha_m t_c \tag{6.29}$$

$$\sigma_{fy} = \left(\sigma_{fy}\right)_{\substack{x=0 \\ y=0}}$$

$$= -\mu_y \frac{G_{cy}}{t_f} \sum_{n=1,3,\dots}^{\infty} D_n \sinh \frac{1}{2} \mu_y \beta_n t_c \qquad (6.30)$$

$$q_x = \left(\tau_{xz}\right)_{\substack{x=a \\ y=0 \\ z=t_c/2}}$$

$$= G_{cx} \sum_{m=1,3,\dots}^{\infty} (-1)^{(m-1)/2} \mu_x \alpha_m C_m \sinh \frac{1}{2} \mu_x \alpha_m t_c \qquad (6.31)$$

$$q_y = \left(\tau_{yz}\right)_{\substack{x=0 \\ y=b \\ z=t_c/2}}$$

$$= G_{cy} \sum_{n=1,3,\dots}^{\infty} (-1)^{(n-1)/2} \mu_y \beta_n D_n \sinh \frac{1}{2} \mu_y \beta_n t_c \qquad (6.32)$$

in which

$\sigma_{fx}, \sigma_{fy}$ = normal stresses in facings at the plate corner;

$q_x, q_y$ = transverse shear stresses at the plate edges.

In the particular case of isotropic core material:

$$E_c = E_{cx} = E_{cy}$$
$$G_c = G_{cx} = G_{cy}$$
$$\mu = \mu_x = \mu_y = (E_c/G_c)^{1/2}$$
$$\psi = \psi' = \mu G_c/E_f t_f$$

and consequently Equs. (6.9) becomes:

$$C_m = \frac{2\,(-1)^{(m-1)/2}}{a\,\alpha_m^2}\;\frac{(\nu\,\Psi\,\xi' - \Delta)}{\frac{\Psi}{\alpha_m}\sinh\frac{1}{2}\mu\,\alpha_m\,t_c + \cosh\frac{1}{2}\mu\,\alpha_m\,t_c}$$

$$D_n = \frac{2\,(-1)^{(n-1)/2}}{b\,\beta_n^2}\;\frac{(\nu\,\Psi\,\xi - \Delta)}{\frac{\Psi}{\beta_n}\sinh\frac{1}{2}\mu\,\beta_n\,t_c + \cosh\frac{1}{2}\mu\,\beta_n\,t_c}$$

With these expressions, Equs. (6.29) to (6.32) can be written in abbreviated form as:

$$\sigma_f = \sigma_{fx} = \sigma_{fy}$$

$$= -\frac{4}{\pi^2}\,\Delta\,\frac{t_c}{t_f}\,(E_c\,G_c)^{1/2}\,K_f \tag{6.33}$$

$$\tau = \tau_{yz}$$

$$= -\frac{5.2}{\pi}\,\Delta\,(E_c\,G_c)^{1/2}\,K_{c1} \tag{6.34}$$

in which

$\sigma_f$  =   maximum normal stresses in facings;

$\tau$   =   maximum interlayer shear stresses.

$$K_f = \sum_{m=1,3,\ldots}^{\infty} \frac{(-1)^{(m-1)/2}}{m^2} \cdot \frac{K_2}{\frac{2}{m\,\pi} \cdot \frac{K_2}{K_1} + \coth \mu \; \frac{m\,\pi}{2\,K_2}} \cdot \frac{\theta_f' + 1}{\theta_f' \, \theta_f - 1}$$

$$K_{c1} = \sum_{n=1,3,\ldots}^{\infty} \frac{1}{n} \; \frac{1}{\frac{2}{n\,\pi\,R} \frac{K_2}{K_1} + \coth \mu \; \frac{n\,\pi\,R}{2\,K_2}}$$

where

$$K_1 = \frac{2\,E_f\,t_f}{t_c\,(E_c\,G_c)^{1/2}}$$

$$K_2 = \frac{2a}{t_c}$$

$$R = \frac{a}{b}$$

$$\theta_f' = \frac{8\,\nu}{\pi^2} \sum_{n=1,3,\ldots}^{\infty} \frac{(-1)^{(n-1)/2}}{n^2} \; \frac{K_2}{R\,K_1} \; \frac{1}{\frac{2}{n\,\pi\,R} \frac{K_2}{K_1} + \coth \mu \; \frac{n\,\pi\,R}{2\,K_2}}$$

$$\theta_f = \frac{8\,\nu}{\pi^2} \sum_{m=1,3,\ldots}^{\infty} \frac{(-1)^{(m-1)/2}}{m^2} \; \frac{K_2}{K_1} \; \frac{1}{\frac{2}{m\,\pi} \frac{K_2}{K_1} + \coth \mu \; \frac{m\,\pi}{2\,K_2}}$$

Numerical values for the factors in Eqs. (6.33) and (6.34) are calculated for the following range of $K_2$, $K_1$, $\mu$ and R.

$K_2$   changes from 10 to 100;
$K_1$   changes from 0.025 to 0.25;

μ      takes the value 1. or 1.5;

R      changes from 1. to 2.

The results are presented in Figs. 6.4 to 6.10.

## 6.5   MOISTURE EFFECT IN COMPOSITE PANELS

To determine the maximum normal stresses in the faces of a panel due to a moisture content change an approximate analysis based on engineering mechanics is conducted, from which simple formulas are derived. Consider the panel in Figs. 6.1. A moisture change in its core induces $S_1$ and $S_2$ dimensional change percent at its upper and lower faces, respectively. To simplify the following analysis, two particular cases are considered: (i) the symmetrical case in which the panel exposed to a uniform moisture change in the core, and (ii) the antisymmetric case as shown in Fig. 6.1. The general case can be obtained by combining these two cases.

### 6.5.1   PANEL SUBJECTED TO A UNIFORM DIMENSIONAL CHANGE

In this case, the moisture-induced strains at the facings center are determined in accordance to Hooke's law as:

$$\varepsilon_{fx} = \frac{1}{E_f} \left( \sigma_{fx} - \nu \, \sigma_{fy} \right)$$

$$\varepsilon_{fy} = \frac{1}{E_f} \left( \sigma_{fy} - \nu \, \sigma_{fx} \right) \tag{6.35}$$

It should be noted that the moisture change produces stresses in the panel in an analogous manner to that in which the temperature affects it. That is, when a plate is exposed to a temperature change at its surfaces, the compatibility condition of the displacement at the interfaces will restrict the free state deformations and hence induce thermal stresses. At the same time, because the free expansion of the faces is in some practical applications bigger than that of the core, it is understood that the core is restricting the faces. In the case of a moisture change in the core, the outer layers are the constraints.

Based on numerical values for the thermal stresses obtained in the preceding sections, it is found that normal stresses in facings at the plate center are equal to X- and Y- directions. Thus, Equ. (6.35) becomes:

$$\varepsilon_f = \varepsilon_{fx} = \varepsilon_{fy}$$

$$= \frac{(1 - \nu)}{E_f} \sigma_f \qquad (6.36)$$

in which

$$\sigma_f = (\sigma_{fx})_{\substack{x=0 \\ y=0}} = (\sigma_{fy})_{\substack{x=0 \\ y=0}}$$

$$\varepsilon_f = \text{normal strain in facings at } x=y=0.$$

In a similar manner

$$\varepsilon_c = \frac{\sigma_c}{E_c}$$

in which

$$\varepsilon_c = \text{normal strain in the core at } x = y = 0.$$

The equilibrium equation of an element at the panel center is

$$2\,\sigma_f\, t_f + \sigma_c\, t_c = 0$$

from which

$$\sigma_c = -2\,\sigma_f\, t_f / t_c$$

and consequently

$$\varepsilon_c = -2\,\sigma_f\, t_f / E_c\, t_c$$

By proper substitution of $\varepsilon_f$ and $\varepsilon_c$ in the compatibility equation of strains at the interfaces between the core and the faces, which is

$$\varepsilon_f - \varepsilon_c = \Delta_m \tag{6.37}$$

it is found that:

$$\sigma_f = \frac{E_f \Delta_m}{1 - \nu + \frac{2 E_f t_f}{E_c t_c}} \tag{6.38}$$

in which

$\Delta_m$ = the relative free strain at the interface between facings and core due to a uniform moisture change.

### 6.5.2 PANEL SUBJECTED TO AN ANTISYMMETRIC DIMENSIONAL CHANGE ACROSS ITS THICKNESS

In this case, the normal strain in the core at $x = y = 0$ can be obtained from

$$\varepsilon_c = z/\rho$$

in which

$\rho$ = the radius of curvature of the core at $x = y = 0$.
$z$ = vertical coordinate in z- direction (Fig. 6.1).

The equilibrium equation of moments on an element in this case is:

$$M_f + M_c = 0$$

in which

$M_f, M_c$ = the moment contributed by facings and core, respectively.

The moments $M_f$ and $M_c$ can be expressed as:

$$M_f = \sigma_f t_f h$$

$$M_c = E_c I_c / \rho, \text{ where } I_c = t_c^3/12$$

Thus, from the equilibrium equation, it is found that:

$$\frac{1}{\rho} = -12 \frac{\sigma_f t_f h}{E_c t_c^3}$$

in which

$$h = t_c + t_f$$

and consequently

$$(\epsilon_c)_{z=t_c/2} = -6 \frac{\sigma_f t_f h}{E_c t_c^2}$$

Substituting this equation and Equ. (6.36) into the compatibility equation (6.37) gives:

$$\sigma_f = \frac{E_f \Delta_m}{1 - \nu + \dfrac{6 E_f t_f h}{E_c t_c^2}} \tag{6.39}$$

It should be noted that Equs. (6.38) and (6.39) can be used to determine the maximum normal stress in the faces of a plate subjected to a uniform temperature change and thermal gradient, respectively. In this case $\Delta$ should be used instead of $\Delta_m$.

## 6.6   NUMERICAL RESULTS

Consider a plate made of two aluminum faces and a wood core, as shown in Fig. 6.11. The plate has the following properties

$t_f$   =   0.025 in.
$t_c$   =   1.74 in.
$2a$   =   36. in.
$2b$   =   24. in.
$E_f$   =   $10^7$ psi.
$\upsilon$   =   0.33
$E_c$   =   $1.2 \times 10^6$ psi.
$G_c$   =   $.16 \times 10^6$ psi.
$\alpha_f$   =   $13 \times 10^{-6}$ in./in. per °F
$\alpha_c$   =   $3 \times 10^{-6}$ in./in. per °F

The plate is subjected to a uniform temperature change of 50°F.

The normal stress distributions in the facings along lines parallel to x and y axes are obtained using the finite element method described earlier.  Because of the symmetry in the plate geometry and temperature distribution, only one-eighth of the plate is considered.  The finite element mesh for this portion of the plate is shown in Fig. 6.12.  The normal stress, $\sigma_f$, is calculated from Equ. (6.23) and the results are presented in Fig. 6.13.

The normal stress distributions are also determined from the analytic solution in Equs. (6.6) and (6.7), (Fig. 6.13).  The discrepancies between the results are due to the coarse size of the finite element mesh used.  However, the accuracy improved when a finer mesh is used as will be shown later.

Another example solved is a panel having the following properties (Fig. 6.14(a))

$t_f$   =   0.025 in.
$t_c$   =   1.74 in.
$2a$   =   36. in.
$E_f$   =   $10^7$ psi.
$\upsilon$   =   0.33
$E_c$   =   $1.2 \times 10^6$ psi.

$G_c$ = .16 x $10^6$ psi.

$\alpha_f$ = 13 x $10^{-6}$ in./in. per °F

$\alpha_c$ = 3 x $10^{-6}$ in./in. per °F

The panel is subjected to a uniform temperature change of 50°F. Because of the symmetry in the geometry and temperature distribution only one quarter is considered in the analysis as shown in Fig. 6.14(b).

The normal and interlayer shear stress distributions along half the strip length are determined using the finite element method (Equ. (6.23)), and the analytic solution in Equs. (6.6) and (6.4). The results are presented in Fig. 6.15. It is seen that the results of the two methods are in good agreement.

As an indication of the correctness of the approximate formula (6.38), the maximum normal stress in the facings of the plate in Fig. 6.13 and in the facings of the strip in Fig. 6.14(a) are determined. The results are shown in Figs. 6.13 and 6.15. It is seen that both the analytic and approximate solutions are in excellent agreement.

## 6.7 REFERENCES

[1] Abel, J.F., and Popov, E.P., "Static and Dynamic Finite Element Analysis of Sandwich Structures", Proc. of the Conference on Matrix Methods in Structural Mechanics, TR-68-150, 1968, Air Force Flight Dynamics Lab., Wright-Patterson Air Force Base, Ohio.

[2] Argyris, J.H., and Fried, I., "The Lumina Element for the Matrix Displacement Method", The Aeronautical Journal of the Royal Aeronautical Society, Vol. 72, June, 1968.

[3] Argyris, J.H., Fried, I., and Scharpf, D.W., "The Hermes 8 Element for the Matrix Displacement Method", The Aeronautical Journal of the Royal Aeronautical Society, Vol. 72, July, 1968.

[4] Bijlaard, P.P., "Thermal Stresses and Deflections in Rectangular Sandwich Plates", J. of the Aerospace Sciences, Vol. 26, No. 4, April, 1959.

[5] Boley, B.A., and Weiner, J.H., Theory of Thermal Stresses, John Wiley and Sons, Inc., 1960.

[6] Chan, H.C., and Cheung, Y.K., "Static and Dynamic Analysis of Multi-Layered Sandwich Plates", Int. J. Mech. Sci., Vol. 14, 1972.

[7]   Chang, C.C., and Ebcioglu, I.K., "Thermoelastic Behaviour of a Simply Supported Sandwich Panel Under Large Temperature Gradient and Edge Compression", J. of the Aerospace Sciences, Vol. 28, No. 6, June, 1961.

[8]   Chong, K.P., Engen, K.O., and Hartsock, J.A., "Thermal Stresses and Deflection of Sandwich Panels", J. Struct. Div., ASCE, Vol. 103, No. ST1, January, 1977.

[9]   Cook, R.D., "Some Elements for Analysis of Plate Bending", J. Eng. Mech. Div., ASCE, Vol. 98, No. EM6, December, 1972.

[10]  Cook, R.D., Concepts and Applications of Finite Element Analysis - A Treatment of the Finite Element Method as used for the Analysis of Displacement, Strain, and Stress, John Wiley and Sons, Inc., 1974.

[11]  Ebcioglu, I.K., "Thermo-Elastic Equations for a Sandwich Panel under Arbitrary Temperature Distribution, Transverse Load, and Edge Compression", Proc. of the 4th U.S. National Congress of Applied Mechanics (American Society of Mechanical Engineers), New York, 1962.

[12]  Gallagher, H.R., Finite Element Analysis Fundamentals, Prentice-Hall, Inc., New Jersey, 1975.

[13]  Gatewood, B.E., Thermal Stresses - With Applications to Airplanes, Missiles, Turbines, and Nuclear Reactors, McGraw-Hill Book Company, Inc. 1957.

[14]  Ha, H.K., "Analysis of Three-Dimensional Orthotropic Sandwich Plate Structures by Finite Element Method", D. Eng. Thesis, Sir George Williams University, Montreal, Canada, 1972.

[15]  Hartsock, J.A., and Chong, K.P., "Analysis of Sandwich Panels with Formed Faces", J. Struct. Div., ASCE, Vol. 102, No. ST4, April, 1976.

[16]  Hoa, S.V., "Hybrid Finite Element Formulation for Thermal Stress Analysis of Plate and Shallow Shell Structures", M.Sc. Thesis, Department of Mechanical Engineering, University of Toronto, Toronto, 1973.

[17]  Holland, I., and Bell, K., Finite Element Method in Stress Analysis, Tapir, the Technical University of Norway, 1969.

[18]  Hussein, R., "Thermal Stress in Flat Metal-Faced Sandwich Panels", M.Eng. Thesis, Centre for Building Studies, Concordia University, Montreal, 1978.

[19]  Isakson, G., and Levy, A., "Finite Element Analysis of Inter-Laminar Shear in Fibrous Composites", J. Composite Materials, Vol. 5, April, 1971.

[20]  Jerzy, P., "Thermal Deformations and Stresses in Rectangular Sandwich Panels

with Non-Rigid Cores", Building Science, Vol. 4, 1969.

[21] Khatua, T.P., and Cheung, U.K., "Triangular Element for Multi-Layer Sandwich Plates", J. Eng. Mech. Div., ASCE, Vol. 98, No. EM5, October, 1972.

[22] Kolar, V., and Nemec, I., "The Efficient Finite Element Analysis of Rectangular and Skew Laminated Plates", International Journal for Numerical Methods in Engineering, Vol. 7, 1973.

[23] Marsh, C., and Nakhla, M., "Thermal Stresses in Metal-Faced Sandwich Panels", Proc. of the Sixth Canadian Congress of Applied Mechanics, Vancouver, May 29 - June 3, 1977.

[24] Marsh, C., and Nakhla, M., "Thermal Stresses in Framed Aluminum Sandwich Panels", ASCE Fall Convention and Exhibit, San Fransisco, October 17-21, 1977.

[25] Mall, S.T., Tong, P., and Pian, T.H.H., "Finite Element Solutions for Laminated Thick Plates", J. Composite Materials, Vol. 6, April, 1972.

[26] Mawenya, A.S., and Davies, J.D., "Finite Element Bending Analysis of Multilayer Plates", International Journal for Numerical Methods in Engineering, Vol. 8, 1974.

[27] Melosh, R.J., "Structural Analysis of Solids", J. Struct. Div., ASCE, Vol. 89, No. ST4, August, 1963.

[28] Monforton, G.R., and Schmit, L.A., "Finite Element Analysis of Sandwich Plates and Cylindrical Shells with Laminated Faces", Proc. of the Conference on Matrix Methods in Structural Mechanics, TR-68-150, 1968, Air Force Flight Dynamics Lab., Wrigth-Patterson Air Force Base, Ohio.

[29] Monforton, G.R., and Michail, M.G., "Finite Element Analysis of Skew Sandwich Plates", Proc. of the American Society of Civil Engineers, Vol. 98, No. EM3, June, 1972.

[30] Newton, R.E., "Degeneration of Brick-Type-Isoparametric Elements", International Journal for Numerical Methods in Engineering, Vol. 8, No. 1, 1974.

[31] Spilker, R.L., "Alternate Hybrid-Stress Elements for Analysis of Multilayer Composite Plates", J. Composite Materials, Vol. 11, January, 1977.

[32] Szilar, R., Theory and Analysis of Plates - Classical and Numerical Methods, Prentice-Hall, Inc., 1974.

[33] Wilson, E.L., Taylor, R.L., Doherty, W.P., and Ghaboussi, J., "Incompatible Displacement Models", Numerical and Computer Methods in Structural Mechanics, Proc. of a Symposium held at Urbana, Illinois, September, 1971.

[34] Wilson, E.L., "The Static Condensation Algorithm", International Journal of Numerical Methods in Engineering, Vol. 8, No. 1, 1974.

[35] Zienkiewicz, O.C., The Finite Element Method in Engineering Sciences, 2nd Edition, McGraw-Hill Book Company, 1971.

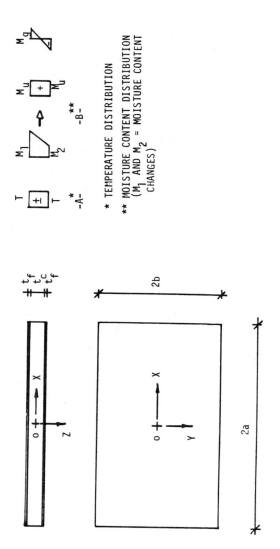

Fig. 6.1 COMPOSITE PANEL UNDER HYGROTHERMAL EFFECT

* TEMPERATURE DISTRIBUTION

** MOISTURE CONTENT DISTRIBUTION
($M_1$ AND $M_2$ = MOISTURE CONTENT
CHANGES)

205

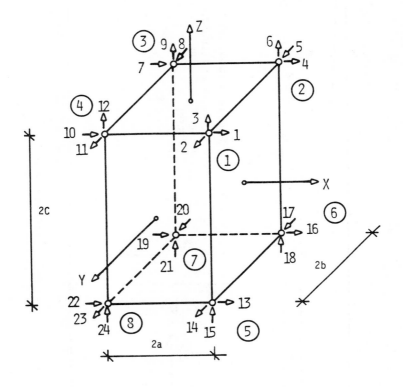

Fig. 6.2 - THE RECTANGULAR PRISM FINITE ELEMENT

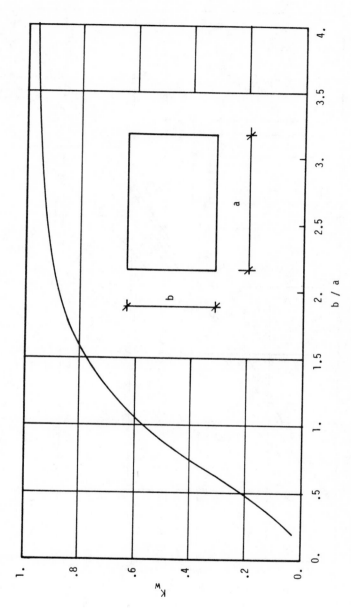

Fig.6.3 – VALUES FOR $K_w$ IN EQU.(6.27)

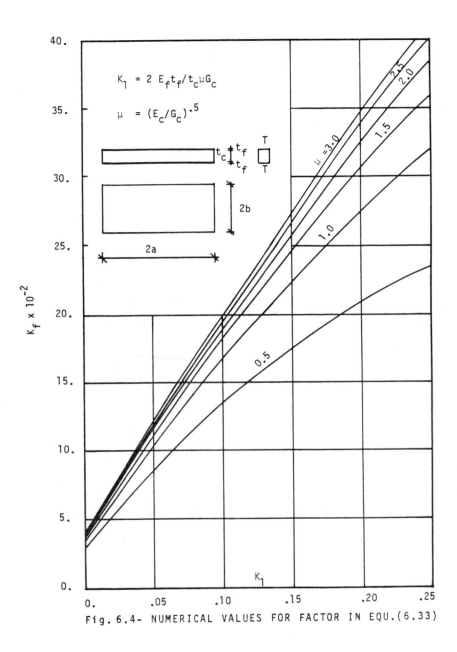

Fig. 6.4- NUMERICAL VALUES FOR FACTOR IN EQU.(6.33)

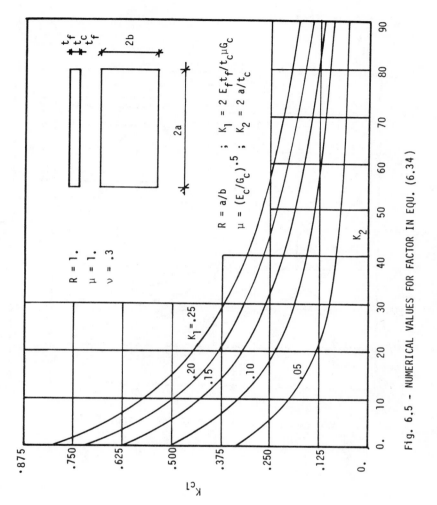

Fig. 6.5 - NUMERICAL VALUES FOR FACTOR IN EQU. (6.34)

209

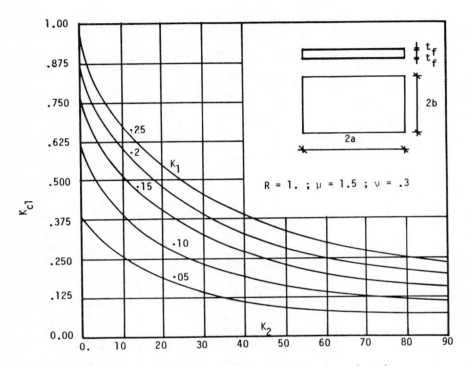

Fig. 6.6 - NUMERICAL VALUES FOR FACTOR IN EQU. (6.34)

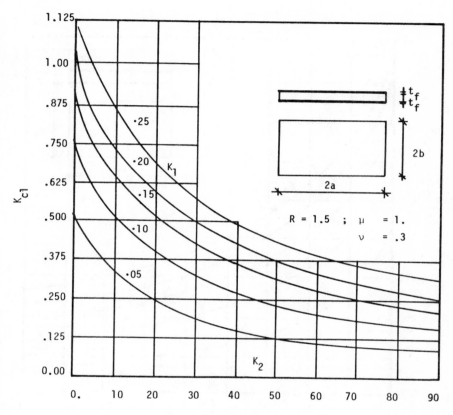

Fig. 6.7 - NUMERICAL VALUES FOR FACTOR IN EQU. (6.34)

211

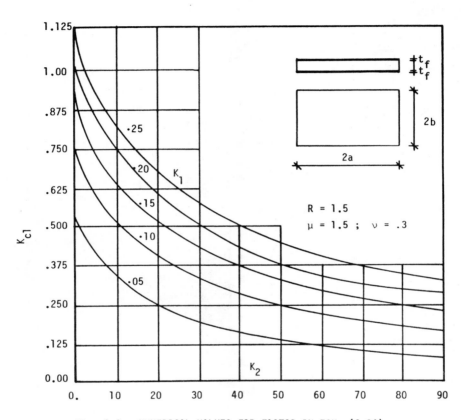

Fig. 6.8 - NUMERICAL VALUES FOR FACTOR IN EQU. (6.34)

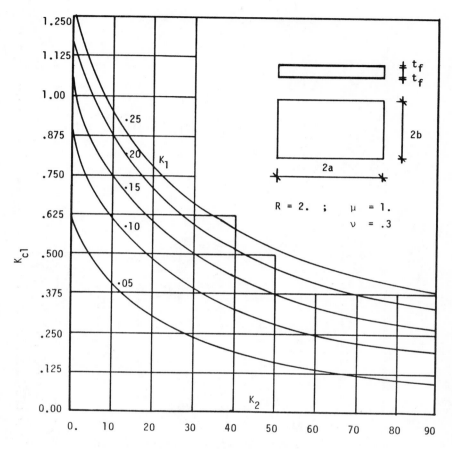

Fig. 6.9 - NUMERICAL VALUES FOR FACTOR IN EQU. (6.34)

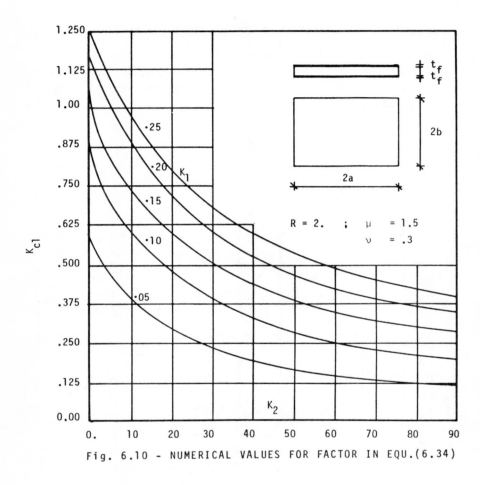

Fig. 6.10 - NUMERICAL VALUES FOR FACTOR IN EQU.(6.34)

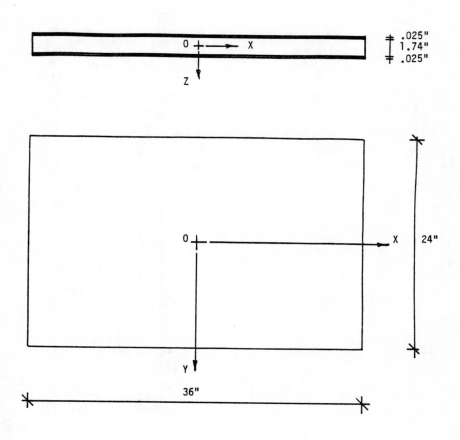

Fig. 6.11  COMPOSITE PLATE USED IN THE FINITE ELEMENT
ANALYSIS OF THERMAL STRESSES

**215**

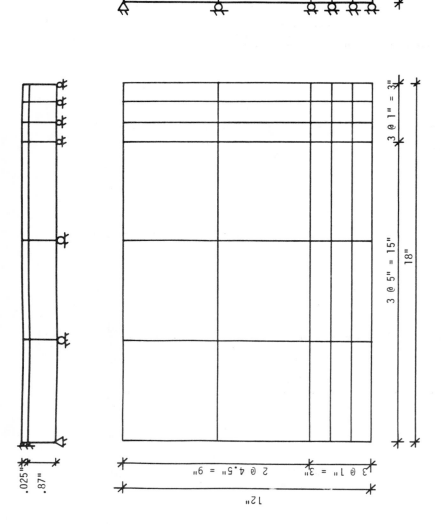

Fig. 6.12   FINITE ELEMENT MESH FOR THE COMPOSITE PLATE IN Fig. 6.11

216

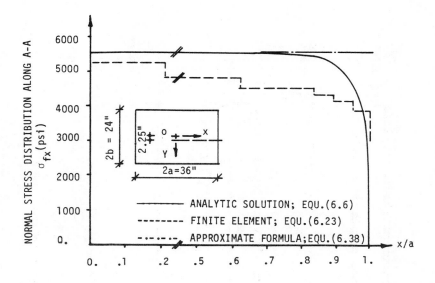

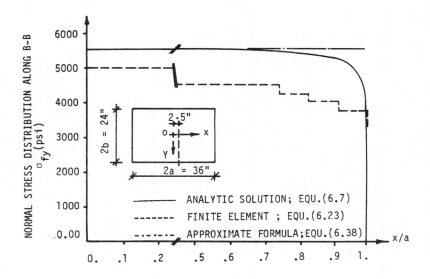

Fig. 6.13 NORMAL STRESS DISTRIBUTION IN FACINGS OF A COMPOSITE PLATE
UNDER UNIFORM TEMPERATURE CHANGE

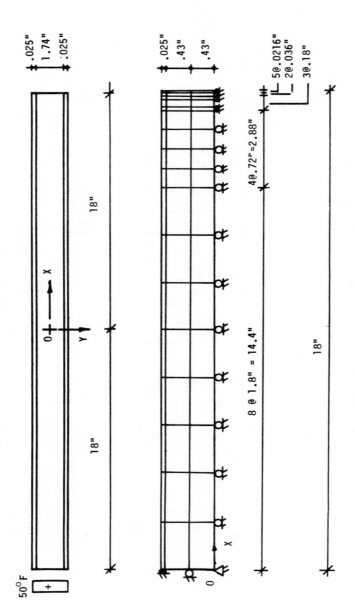

Fig. 6.14 COMPOSITE PANEL UNDER UNIFORM TEMPERATURE CHANGE

218

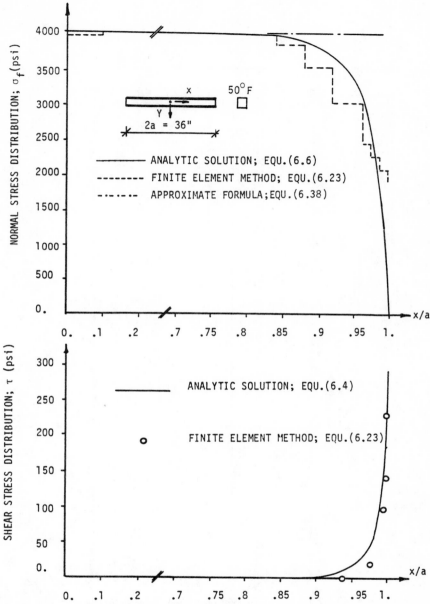

Fig. 6.15 NORMAL AND SHEAR STRESS DISTRIBUTIONS IN COMPOSITE PANEL UNDER UNIFORM TEMPERATURE CHANGE

**219**

# VII

# **Vibrations of Composite Plates**

All loads considered in the preceding chapters have been static. In practice dynamic disturbances induced by applied loads are often encountered. There are many sources for a dynamic load, such as wind gusts, earthquakes, shock or blast, sound and unbalanced operating equipments. The objective of dynamics of structures is usually to determine their response or the dynamic performance. This will be discussed in this chapter with regard to composite plates. Due to the complexity of the subject, a very brief summary is presented first on the fundamental equations of plate theory.

## 7.1 FUNDAMENTAL EQUATIONS OF PLATE THEORY

The differential equation of motion for the transverse displacement w of a plate is given by

$$D\nabla^4 w + \rho(\partial^2 w/\partial t^2) = 0 \qquad (7.1)$$

in which

| | | |
|---|---|---|
| D | = | the flexural rigidity |
| | = | $Eh^3/12(1 - v^2)$; |
| E | = | Young's modulus; |
| h | = | the plate thickness; |
| v | = | Poisson's ratio; |
| $\rho$ | = | mass density per unit area of the plate; |
| t | = | time; |
| $\nabla^4$ | = | $\nabla^2\nabla^2$, where $\nabla^2$ is the Laplacian operator. |

$$(7.2)$$

When free vibrations are assumed, the motion is expressed as

$$w = W \cos\omega\tau \qquad (7.3)$$

where $\omega$ is the circular frequency (expressed in radians/unit time) and W is a function only of the position coordinates.

**221**

Substituting equation (7.3) into equation (7.1) yields:

$$(\nabla^4 - k^4)W = 0 \tag{7.4}$$

where k is a parameter of convenience defined as:

$$k^4 = \rho\omega^2/D \tag{7.5}$$

It is usually convenient to factor equation (7.4) into

$$(\nabla^2 + k^2)(\nabla^2 - k^2)W = 0 \tag{7.6}$$

whence, by the theory of linear differential equations, the complete solution to equation (7.6) can be obtained by superimposing the solutions to the equations

$$\nabla^2 W_1 + k^2 W_1 = 0$$
$$\nabla^2 W_2 - k^2 W_2 = 0 \tag{7.7}$$

The Laplacian operator in rectanular coordinates is

$$\nabla^2 = \partial^2/\partial x^2 + \partial^2/\partial y^2 \tag{7.8}$$

Bending and twisting moments are related to the displacements by

$$
\begin{aligned}
M_x &= -D[(\partial^2 w/\partial x^2) + \nu\,(\partial^2 w/\partial y^2)] \\
M_y &= -D[(\partial^2 w/\partial y^2) + \nu\,(\partial^2 w/\partial x^2)] \\
M_{xy} &= -D(1 - \nu)(\partial^2 w/\partial x \partial y)
\end{aligned} \tag{7.9}
$$

Transverse shearing forces are given by

$$
\begin{aligned}
Q_x &= -D(\partial/\partial x)(\nabla^2 w) \\
Q_y &= -D(\partial/\partial y)(\nabla^2 w)
\end{aligned} \tag{7.10}
$$

and the Kelvin-Kirchhoff edge reations are

$$
\begin{aligned}
V_x &= Q_x + (\partial M_{xy}/\partial y) \\
V_y &= Q_y + (\partial M_{xy}/\partial x)
\end{aligned} \tag{7.11}
$$

General solutions to equation (7.4) in rectangular coordinates may be obtained by assuming Fourier series in one of the variables, say x; that is:

$$W(x,y) = \sum_{m=1}^{\infty} Y_m(y) \sin \alpha x + \sum_{m=0}^{\infty} Y_m^*(y) \cos \alpha x \qquad (7.12)$$

Substituting equation (7.12) into equation (7.7) yields:

$$d^2 Y_{m1}/dy^2 + (k^2 - \alpha^2)Y_{m1} = 0 \qquad (7.13)$$
$$d^2 Y_{m2}/dy^2 + (k^2 + \alpha^2)Y_{m2} = 0$$

and two similar equations for $Y_m^*$. With the assumption that $K^2 > \alpha^2$, solutions to equations (7.13) are well known as:

$$Y_{m1} = A_m \sin (k^2 - \alpha^2 y)^{1/2} + B_m \cos (k^2 - \alpha^2 y)^{1/2} \qquad (7.14)$$
$$Y_{m2} = C_m \sin (k^2 + \alpha^2 y)^{1/2} + D_m \cosh (k^2 + \alpha^2 y)^{1/2}$$

where $A_m, \ldots, D_m$ are arbitrary coefficients determining the mode shape and are obtained from the boundary conditions. If $k^2 < \alpha^2$, it is necessary to rewrite $Y_{m1}$ as:

$$Y_{m1} = A_m \sinh (\alpha^2 - k^2 y)^{1/2} + B_m \cosh (\alpha^2 - k^2 y)^{1/2} \qquad (7.15)$$

Thus the complete solution to equation (7.4) may be written as:

$$W(x,y) = \sum_{m=1}^{\infty} [A_m \sin (k^2 - \alpha^2 y)^{1/2} + B_m \cos (k^2 - \alpha^2 y)^{1/2} + C_m \sinh (k^2 + \alpha^2 y)^{1/2}$$
$$+ D_m \cosh (k^2 + \alpha^2 y)^{1/2}] \sin \alpha x + \sum_{m=0}^{\infty} [A_m^* \sin (k^2 - \alpha^2 y)^{1/2} +$$
$$B_m^* \cos (k^2 - \alpha^2 y)^{1/2} + C_m^* \sinh (k^2 + \alpha^2 y)^{1/2}$$
$$+ D_m^* \cosh (k^2 + \alpha^2 y)^{1/2}] \cos \alpha x \qquad (7.16)$$

## 7.2 NATURAL FREQUENCIES OF PLATES

Warburton [3] presented solutions for the natural frequencies of plates. He used the Rayleigh method with deflection functions considered as:

$$W(x,y) = X(x)Y(y) \qquad (7.17)$$

in which

$X(x), Y(y)$ = fundamental mode shapes of beams having the boundary conditions of the plate.

This choice of functions exactly satisfies all boundary conditions for the plate. The six possible distinct sets of boundary conditions along the edges $x = 0$ and $x = a$ are satisfied by the following mode shapes:

(a) Simply Supported at $x = 0$ and $x = a$:

$$X(x) = \sin(m - 1)\pi x/a \qquad (m = 2, 3, 4, \ldots) \qquad (7.18)$$

(b) Clamped at $x = 0$ and $x = a$:

$$X(x) = \cos\gamma_1[(x/a) - (1/2)] + [\sin(\gamma_1/2)/\sinh(\gamma_1/2)] \cos\gamma_1[(x/a) - (1/2)]$$
$$(m = 2, 4, 6, \ldots) \qquad (7.19)$$

in which

$$\gamma_1 = \text{roots of:} \quad \tan(\gamma_1/2) + \tanh(\gamma_1/2) = 0 \qquad (7.20)$$

and

$$X(x) = \sin\gamma_2\,[(x/a) - (1/2)] - [\sin(\gamma_2/2) / \sinh(\gamma_2/2)] \sinh\gamma_2\,[(x/a) - (1/2)]$$
$$(m = 3, 5, 7, \ldots) \qquad (7.21)$$

in which

$$\gamma_2 = \text{roots of:} \quad \tan(\gamma_2/2) - \tanh(\gamma_2/2) = 0 \qquad (7.22)$$

(c) Free at $x = 0$ and $x = a$:

$$X(x) = 1 \qquad (m = 0) \qquad (7.23)$$
$$X(x) = 1 - 2x/a \qquad (m = 1) \qquad (7.24)$$

$$X(x) = \cos\gamma_1\,[(x/a) - (1/2)] + [\sin(\gamma_1/2) / \sinh(\gamma_1/2)] \cosh\gamma_1\,[(x/a) - (1/2)]$$
$$(m = 2, 4, 6, \ldots) \qquad (7.25)$$

and

$$X(x) = \sin \gamma_2 [(x/a) - (1/2)] + [\sin (\gamma_2/2) / \sinh (\gamma_2/2)] \sinh\gamma_2 [(x/a) - (1/2)]$$
$$(m = 3, 5, 7, \ldots) \tag{7.26}$$

with $\gamma_1$ and $\gamma_2$ as defined in equations (7.20) and (7.22).

(d)  Clamped at $x = 0$ and Free at $x = a$:

$$X(x) = \cos (\gamma_3 x/a) - \cosh (\gamma_3 x/a)$$
$$+ [(\sin\gamma_3 - \sinh\gamma_3) / (\cos\gamma_3 - \cosh\gamma_3)] [\sin (\gamma_3 x/a) - \sinh (\gamma_3 x/a)]$$
$$(m = 1, 2, 3, \ldots) \tag{7.27}$$

where

$$\cos\gamma_3 \cosh\gamma_3 = -1 \tag{7.28}$$

(e)  Clamped at $x = 0$ and Simply Supported at $x = a$:

$$X(x) = \sin\gamma_2 [(x/2a) - (1/2)] - [\sin (\gamma_2/2) / \sinh (\gamma_2/2)] \sinh\gamma_2 [(x/2a) - (1/2)]$$
$$(m = 2, 3, 4, \ldots) \tag{7.29}$$

with $\gamma_2$ as defined in equation (7.22).

(f)  Free at $x = 0$ and Simply Supported at $x = a$:

$$X(x) = 1 - (x/a) \qquad (m = 1) \tag{7.30}$$

$$X(x) = \sin\gamma_2 [(x/2a) - (1/2)] + [\sin (\gamma_2/2) / \sinh (\gamma_2/2)] \sinh\gamma_2 [(x/2a) - (1/2)]$$
$$(m = 2, 3, 4, \ldots) \tag{7.31}$$

with $\gamma_2$ as defined in equation (7.22).

The functions $Y(y)$ are similarly chosen by the conditions at $y = 0$ and $y = a$ by replacing x by y, a by b, and m by n in equations (7.18) to (7.31).  The indicators n and m are seen to be the number of nodal lines lying in the x- and y- directions, respectively, including the boundaries as nodal lines, except when the boundary is free.

The frequency $\omega$ is obtained as:

$$\omega^2 = \pi^4 D/a^4 \rho \qquad \{G_x^4 + G_y^4\,(a/b)^4 + 2(a/b)^2\,[\nu\,H_x H_y + (1-\nu)\,J_x J_y\,]\} \qquad (7.32)$$

where $G_x$, $H_x$, and $J_x$ are functions determined from table 7.1 according to the conditions at $x = 0$ and $x = a$. The quantities $G_y$, $H_y$, and $J_y$ are obtained from table 7.1 by replacing x by y and m by n.

TABLE 7.1 - Frequency Coefficients in Equation (7.32) [3]

| Boundary Conditions at | m | $G_x$ | $H_x$ | $J_x$ |
|---|---|---|---|---|
| $SS_a$ $SS_b$ | 2, 3, 4, . . . | m - 1 | $(m-1)^2$ | $(m-1)^2$ |
| $C_a$ | 2 | 1.506 | 1.248 | 1.248 |
| $C_b$ | 3, 4, 5, . . . | m - j | AE | AE |
| $F_a$ | 0 | 0 | 0 | 0 |
|  | 1 | 0 | 0 | $12/\pi^2$ |
|  | 2 | 1.506 | 1.248 | 5.017 |
| $F_b$ | 3, 4, 5, . . . | m - j | AE | $A\,[1 + (6/(m-j)\,\pi)]$ |
| $C_a$ $SS_b$ | 2, 3, 4, . . . | m - (3/4) | BD | BD |
| $F_a$ | 1 | 0 | 0 | $3/\pi^2$ |
| $SS_b$ | 2, 3, 4, . . . | m - (3/4) | BD | $B\,[1 + (3/(m - (3/4))\,\pi)]$ |
| $C_a$ | 1 | 0.597 | -0.0870 | 0.471 |
|  | 2 | 1.494 | 1.347 | 3.284 |
| $F_b$ | 3, 4, 5, . . . | m - j | AE | $A\,[1 + (2/(m-j)\,\pi)]$ |

$a_x$ = 0.
$b_x$ = a.
A = $(m - j)^2$
B = $[m - (3/4)]^2$
C = Clamped.
D = $[1 - (1/(m - (3/4))\pi)]$
E = $[1 - (2/(m - j)\pi)]$
F = Free.
SS = Simply Supported.

Should the plate be an isotropic one, the differential equation for the transverse bending is given by:

$$D_x (\partial^4 w/\partial x^4) + 2D_{xy} (\partial^4 w/\partial x^2 \partial y^2) + D_y (\partial^4 w/\partial y^4) + \rho (\partial^2 w/\partial t^2) = 0 \qquad (7.33)$$

The moment-curvature relations are:

$$
\begin{aligned}
M_x &= -D_x [(\partial^2 w/\partial x^2) + v_y (\partial^2 w/\partial y^2)] \\
M_y &= -D_y [(\partial^2 w/\partial y^2) + v_x (\partial^2 w/\partial x^2)] \\
M_{xy} &= -2D_k (\partial^2 w/\partial x \partial y)
\end{aligned}
\qquad (7.34)
$$

If the orthotropic constants $D_x'$, $D_y'$, and $D_{xy}'$ are known with respect to the x' and y' coordinate axes, the orthotropic constants $D_x$, $D_y$, and $D_{xy}$ can be determined from:

$$
\begin{aligned}
D_x &= D_x' \cos^4\phi + D_y' \sin^4\phi + 2D_{xy}' \sin^2\phi \cos^2\phi \\
D_y &= D_x' \sin^4\phi + D_y' \cos^4\phi + 2D_{xy}' \sin^2\phi \cos^2\phi \\
D_{xy} &= (3D_x' + 3D_y' - 2D_{xy}') \sin^2\phi \cos^2\phi + D_{xy}' (\cos^2\phi - \sin^2\phi)^2
\end{aligned}
\qquad (7.35)
$$

Hearmon [2] used the Rayleigh method to determine the frequencies of rectangular orthotropic plates. He found that:

$$\omega^2 = (1/\rho) [(A^4 D_x / a^4) + (B^4 D_y / b^4) + (2CD_{xy} / a^2 b^2)] \qquad (7.36)$$

in which A, B, and C are summarized in table 7.2 for the various boundary conditions and modes. The terms $\gamma_i$ and $\varepsilon_i$ in table 7.2 are given by

$$
\begin{aligned}
\gamma_0 &= m \pi \\
\gamma_1 &= [m + (1/4)] \pi
\end{aligned}
$$

$$\gamma_2 = [m + (1/2)]\,\pi$$
$$\varepsilon_0 = n\,\pi$$
$$\varepsilon_1 = [n + (1/4)]\,\pi$$
$$\varepsilon_2 = [n + (1/2)]\,\pi$$

TABLE 7.2 - Frequency Coefficients in Equations (7.36) [2]

| Boundary Conditions | A | B | C | m | n |
|---|---|---|---|---|---|
| C-C-C-C | 4.730 | 4.730 | 151.3 | 1 | 1 |
| | 4.730 | $\varepsilon_2$ | $12.30\,\varepsilon_2\,(\varepsilon_2-2)$ | 1 | 2, 3, 4, ... |
| | $\gamma_2$ | 4.730 | $12.30\,\gamma_2\,(\gamma_2-2)$ | 2, 3, 4, ... | 1 |
| | $\gamma_2$ | $\varepsilon_2$ | $\gamma_2\,\varepsilon_2\,(\gamma_2-2)(\varepsilon_2-2)$ | 2, 3, 4, ... | 2, 3, 4, ... |
| C-C-C-SS | 4.730 | $\varepsilon_1$ | $12.30\,\varepsilon_1\,(\varepsilon_1-1)$ | 1, 2, 3, ... | 1 |
| | $\gamma_2$ | $\varepsilon_1$ | $\gamma_2\,\varepsilon_1\,(\gamma_2-2)(\varepsilon_1-1)$ | 1, 2, 3, ... | 2, 3, 4, ... |
| C-SS-C-SS | 4.730 | $\varepsilon_0$ | $12.30\,\varepsilon_0^2$ | 1 | 1, 2, 3, ... |
| | $\gamma_2$ | $\varepsilon_0$ | $\gamma_2\,\varepsilon_0^2\,(\gamma_2-2)$ | 2, 3, 4, ... | 1, 2, 3, ... |
| C-SS-SS-C | $\gamma_1$ | $\varepsilon_1$ | $\gamma_1\,\varepsilon_1\,(\gamma_1-1)(\varepsilon_1-1)$ | 1, 2, 3, ... | 1, 2, 3, ... |
| C-SS-SS-SS | $\gamma_1$ | $\varepsilon_0$ | $\gamma_1\,\varepsilon_0^2\,(\gamma_1-1)$ | 1, 2, 3, ... | 1, 2, 3, ... |
| SS-SS-SS-SS | $\gamma_0$ | $\varepsilon_0$ | $\gamma_0^2\,\varepsilon_0^2$ | 1, 2, 3, ... | 1, 2, 3, ... |

SS = Simply Supported.
C = Clamped.

## 7.3   FREQUENCY - RESPONSE OF ORTHOTROPIC COMPOSITE PLATES

Consider now a composite plate, the equations of motion are given by [3]:

$$(\partial M_x/\partial x) + (\partial M_{xy}/\partial y) - Q_x = I\ (\partial^2\phi_x/\partial t^2)$$
$$(\partial M_{xy}/\partial x) + (\partial M_y/\partial y) - Q_y = I\ (\partial^2\phi_y/\partial t^2) \qquad (7.37)$$
$$(\partial Q_x/\partial x) + (\partial Q_y/\partial y) + p = M\ (\partial^2 w/\partial t^2)$$

in which

| | | |
|---|---|---|
| $M_x, M_y, M_{xy}$ | = | bending and torsional moments, respectively; |
| $Q_x, Q_y$ | = | transverse shear forces; |
| w | = | lateral deflection; |
| $\phi_x, \phi_y$ | = | slopes in the x-z and y-z planes, respectively; |
| I | = | mass moment of inertia with respect to the middle plane per unit area; |
| | = | $(1/12)\,\rho_c\,t_c^3 + (1/2)\,\rho_f\,t_f\,t_c^2$; |
| M | = | mass of the plate per unit area; |
| | = | $\rho_c\,t_c + 2\rho_f\,t_f$; |
| p | = | applied load; |
| $t_c, t_f$ | = | thickness of core and face, respectively; |
| $\rho$ | = | density; |
| a, b | = | dimensions parallel to x- and y- axes, respectively. |

It is known that the relations between the resisting forces and the displacement components are:

$$Q_x = S_x\,[\phi_x + (\partial w/\partial x)]$$
$$Q_y = S_y\,[\phi_y + (\partial w/\partial y)]$$
$$M_x = D_x\,(\partial\phi_x/\partial x) + D_{xy}\,(\partial\phi_y/\partial y) \qquad (7.38)$$
$$M_y = D_y\,(\partial\phi_y/\partial y) + D_{xy}\,(\partial\phi_x/\partial x)$$
$$M_{xy} = S_{xy}\,[(\partial\phi_x/\partial y) + (\partial\phi_y/\partial x)]$$

in which

$$S_x = G_{xc}\,t_c;$$
$$S_y = G_{yc}\,t_c;$$

$$S_{xy} = (1/12)[G_{zc} t_c^2 + 6G_{zf} t_c^2 t_f];$$
$$D_x = (1/12)[E_{xc} t_c^3 + 6E_{xc} t_c^2 t_f];$$
$$D_y = (1/12)[E_{yc} t_c^3 + 6E_{yf} t_c^2 t_f];$$
$$D_{xy} = (1/12)[E_{xyc} t_c^3 + 6E_{xyf} t_c^2 t_f];$$

$E$ = modulus of elasticity;
$G$ = shear modulus.

The stress-strain relations for the core are:

$$\sigma_{yz} = 2G_{xc} \varepsilon_{yz}$$
$$\sigma_{xz} = 2G_{yc} \varepsilon_{xz}$$
$$\sigma_{xy} = 2G_{zc} \varepsilon_{xy} \qquad (7.39)$$
$$\sigma_x = E_{xc} \varepsilon_{xc} + E_{xyc} \varepsilon_{yc}$$
$$\sigma_y = E_{yc} \varepsilon_{yc} + E_{xyc} \varepsilon_{xc}$$

Also, the stress-strain relations for the faces are:

$$\sigma_{xy} = 2G_{zf} \varepsilon_{xy}$$
$$\sigma_x = E_{xf} \varepsilon_{xf} + E_{xyf} \varepsilon_{yf} \qquad (7.40)$$
$$\sigma_y = E_{yf} \varepsilon_{yf} + E_{xyf} \varepsilon_{xf}$$

By substituting Eqs. 7.38 into Eqs. 7.37, one can obtain the following equations:

$$D_x' (\partial^2 \phi_x / \partial x^2) + D_{xy} (\partial^2 \phi_y / \partial x \partial y) + S_{xy}' (\partial^2 \phi_y / \partial x \partial y) + S_{xy} (\partial^2 \phi_x / \partial y^2)$$
$$- S_x'[\phi_x + (\partial w/\partial x)] = I \ (\partial^2 \phi_x / \partial t^2)$$
$$D_y' (\partial^2 \phi_y / \partial y^2) + D_{xy} (\partial^2 \phi_x / \partial x \partial y) + S_{xy}' (\partial^2 \phi_x / \partial x \partial y) + S_{xy}' (\partial^2 \phi_y / \partial x^2) \qquad (7.41)$$
$$- S_y' [\phi_y + (\partial w/\partial y)] = I \ (\partial^2 \phi_y / \partial t^2)$$
$$S_x'(\partial \phi_x / \partial x) + S_y' (\partial \phi_y / \partial y) + S_x'(\partial^2 w/\partial x^2) + S_y' (\partial^2 w/\partial y^2) + p = M \ (\partial^2 w/\partial t^2)$$

where the prime denotes complex shear, flexural and twisting rigidities.

To solve equations 7.41, the boundary conditions must be defined.  For a simply supported plate.

$$M_x = \phi_y = w = 0 \qquad\qquad\qquad \text{at } x = 0$$
and $x = a$

$$M_y = \phi_x = w = 0 \qquad \text{at } y = 0$$

and $y = b$ (7.42)

Now, a solution for $\phi_x$, $\phi_y$, and $w$ is considered as:

$$\phi_x = \sum_m \sum_n \phi_{xmn} \cos\alpha_m x \sin\beta_n y \, e^{i\omega t} = \phi_x' \, e^{i\omega t}$$

$$\phi_y = \sum_m \sum_n \phi_{ymn} \sin\alpha_m x \cos\beta_n y \, e^{i\omega t} = \phi_y' \, e^{i\omega t} \qquad (7.43)$$

$$w = \sum_m \sum_n W_{mn} \sin\alpha_m x \sin\beta_n y \, e^{i\omega t} = w' \, e^{i\omega t}$$

in which

$$\alpha_m = m\pi/a$$
$$\beta_n = n\pi/a$$
$$m, n = 1, 2, 3, \ldots$$

By substituting Eqs. 7.43 into Eqs. 7.40, a system of linear equations is found as:

$$(D_x' \alpha_m^2 + S_{xy}' \beta_n^2 + S_x' - Iw^2)\,\phi_{xmn} + (D_{xy}' \alpha_m \beta_n + S_{xy}' \alpha_m \beta_n)\,\phi_{xmn}$$
$$+ S_x' \alpha_m W_{mn} = 0$$
$$(D_{xy}' \alpha_m \beta_n + S_{xy}' \alpha_m \beta_n)\,\phi_{xmn} + (D_y' \beta_n^2 + S_{xy}' \alpha_m^2 + S_y' - Iw^2)\,\phi_{ymn} \qquad (7.44)$$
$$+ S_y' \beta_n W_{mn} = 0$$
$$S_x' \alpha_m \phi_{xmn} + S_y' \beta_n \phi_{ymn} + (S_x' \alpha_m^2 + S_y' \beta_n^2 - Mw^2)\,W_{mn} = P_{mn}$$

in which $P_{mn}$ is the coefficient of Fourier-series expansion of the applied load, i.e.

$$P = \sum_m \sum_n P_{mn} \sin\alpha_m x \sin\beta_n y \, e^{i\omega t}$$

A solution of Eqs. 7.44 is found as:

$$W_{mn} = P_{mn} [(AI\omega^2 + B)/(I^2\omega^4 + CI\omega^2 + D) + E - M\omega^2]^{-1}$$
$$\phi_{xmn} = W_{mn}(S_x' \alpha_m I\omega^2 + F)/(I^2\omega^4 + CI\omega^2 + D) \qquad (7.45)$$
$$\phi_{ymn} = W_{mn}(S_y' \beta_n I\omega^2 + G)/(I^2\omega^4 + CI\omega^2 + D)$$

in which

$$A = S_x'^2 \alpha_m^2 + S_y'^2 \beta_n^2$$

$$B = -S_{xy}' S_x'^2 \alpha_m^4 - S_{xy}' S_y'^2 \beta_n^4 + (-D_y' S_x'^2 + 2D_{xy}' S_x' S_y' + 2S_{xy}' S_x' S_y' + D_x' S_y'^2) \alpha_m^2 \beta_n^2 - S_x'^2 S_y' \alpha_m^2 - S_x' S_y'^2 \beta_n^2$$

$$C = -(D_x' \alpha_m^2 + D_y' \beta_n^2 + S_{xy}' \alpha_m^2 + S_{xy}' \beta_n^2 + S_x' + S_y')$$

$$D = (D_x' \alpha_m^2 + S_{xy}' \beta_n^2 + S_x')(D_y' \beta_n^2 + S_{xy}' \alpha_m^2 + S_y') - (D_{xy}' + S_{xy}')^2 \alpha_m^2 \beta_n^2$$

$$E = S_x' \alpha_m^2 + S_y' \beta_n^2$$

$$F = -S_x' \alpha_m (D_y' \beta_n^2 + S_{xy}' \alpha_m^2 + K_y') + K_y' \beta_n (D_{xy}' \alpha_m \beta_n + S_{xy}' \alpha_m \beta_n)$$

$$G = -S_y' \beta_n (D_x' \alpha_m^2 + S_{xy}' \beta_n^2 + S_x') + S_x' \alpha_m (D_{xy}' \alpha_m \beta_n + S_{xy}' \alpha_m \beta_n)$$

Once the displacements are determined, any quantity of interest, such as stresses, can be evaluated using Eqs. 7.37.

## 7.4   REFERENCES

[1] Bieniek, M.P., and Freudenthal, A.M., "Frequency - Response Functions of Orthotropic Sandwich Plates," Journal of the Aerospace Science, Vol. 28, 1961.

[2] Hearmon, R.F.S., "The Frequency of Flexural Vibration of Rectangular Orthotropic Plates With Clamped or Supported Edges", J. Appl. Mech., Vol. 26, Nos. 3-4, Dec. 1959.

[3] Warburton, G.B., "The Vibration of Rectangular Plates," Proc. Inst. Mech. Engr., Ser. A, Vol. 168, 1954.

# APPENDIX A

# Numerical Values for the Factors in the Practical Formulas for Simply Supported Composite Plates

R = $\dfrac{a}{b}$

ν = .30

| ρ | KWB | KWS | KQX | KQY | KMX | KMY | KMXY |
|---|---|---|---|---|---|---|---|
| 1.0000 | .0041 | .0737 | .3357 | .3357 | .0479 | .0479 | -.0325 |
| 1.1000 | .0033 | .0666 | .3129 | .2958 | .0408 | .0459 | -.0267 |
| 1.2000 | .0027 | .0602 | .2921 | .2621 | .0348 | .0435 | -.0220 |
| 1.3000 | .0022 | .0545 | .2732 | .2333 | .0298 | .0411 | -.0182 |
| 1.4000 | .0018 | .0494 | .2561 | .2067 | .0256 | .0385 | -.0152 |
| 1.5000 | .0015 | .0448 | .2407 | .1875 | .0222 | .0361 | -.0127 |
| 1.6000 | .0013 | .0407 | .2267 | .1690 | .0193 | .0337 | -.0107 |
| 1.7000 | .0011 | .0371 | .2141 | .1530 | .0158 | .0314 | -.0091 |
| 1.8000 | .0009 | .0339 | .2027 | .1369 | .0148 | .0293 | -.0078 |
| 1.9000 | .0007 | .0310 | .1923 | .1266 | .0131 | .0273 | -.0067 |
| 2.0000 | .0006 | .0285 | .1828 | .1158 | .0116 | .0254 | -.0058 |
| 3.0000 | .0002 | .0136 | .1217 | .0545 | .0045 | .0132 | -.0018 |
| 4.0000 | .0001 | .0078 | .0908 | .0310 | .0024 | .0077 | -.0007 |
| 5.0000 | .0000 | .0050 | .0722 | .0199 | .0015 | .0050 | -.0004 |

TABLE A.1. NUMERICAL VALUES FOR THE FACTORS IN EQUS. (2.38), (2.42), and (2.48).

R = $\dfrac{a}{b}$

ν = .30

| R | KWB | KWS | KQX | KQY | KMX | KMY | KMXY |
|---|---|---|---|---|---|---|---|
| 1.0000 | .0020 | .0368 | .1678 | .1678 | .0239 | .0239 | -.0162 |
| 1.1000 | .0017 | .0333 | .1565 | .1479 | .0204 | .0229 | -.0133 |
| 1.2000 | .0014 | .0301 | .1461 | .1310 | .0174 | .0218 | -.0110 |
| 1.3000 | .0011 | .0273 | .1366 | .1167 | .0149 | .0205 | -.0091 |
| 1.4000 | .0009 | .0247 | .1281 | .1044 | .0128 | .0193 | -.0076 |
| 1.5000 | .0008 | .0224 | .1203 | .0937 | .0111 | .0180 | -.0064 |
| 1.6000 | .0006 | .0204 | .1134 | .0845 | .0096 | .0168 | -.0054 |
| 1.7000 | .0005 | .0186 | .1070 | .0765 | .0084 | .0157 | -.0046 |
| 1.8000 | .0004 | .0169 | .1013 | .0695 | .0074 | .0146 | -.0039 |
| 1.9000 | .0004 | .0155 | .0961 | .0633 | .0065 | .0136 | -.0033 |
| 2.0000 | .0003 | .0142 | .0914 | .0579 | .0058 | .0127 | -.0029 |
| 3.0000 | .0001 | .0068 | .0649 | .0273 | .0023 | .0066 | -.0009 |
| 4.0000 | .0000 | .0039 | .0454 | .0155 | .0012 | .0039 | -.0004 |
| 5.0000 | .0000 | .0025 | .0361 | .0100 | .0008 | .0025 | -.0002 |

**TABLE A.2.** NUMERICAL VALUES FOR THE FACTORS IN EQU. (2.53).

$$R = \frac{a}{b}$$

$$\xi' = .17 \quad (\xi' = \xi/a)$$
$$\eta' = .17 \quad (\eta' = \eta/b)$$
$$\nu = .30$$

| R | KWB | KWS | KQX | KQY | KMX | KMY | KMXY |
|---|---|---|---|---|---|---|---|
| 1.0000 | .0001 | .0019 | .0160 | .0160 | .0012 | .0012 | -.0054 |
| 1.1000 | .0001 | .0017 | .0164 | .0128 | .0010 | .0012 | -.0044 |
| 1.2000 | .0001 | .0015 | .0167 | .0104 | .0008 | .0012 | -.0037 |
| 1.3000 | .0001 | .0014 | .0168 | .0084 | .0006 | .0012 | -.0031 |
| 1.4000 | .0001 | .0012 | .0168 | .0069 | .0005 | .0011 | -.0026 |
| 1.5000 | .0000 | .0011 | .0167 | .0058 | .0004 | .0010 | -.0022 |
| 1.6000 | .0000 | .0009 | .0165 | .0048 | .0003 | .0009 | -.0019 |
| 1.7000 | .0000 | .0008 | .0163 | .0040 | .0002 | .0009 | -.0017 |
| 1.8000 | .0000 | .0007 | .0160 | .0034 | .0001 | .0008 | -.0015 |
| 1.9000 | .0000 | .0006 | .0157 | .0029 | .0001 | .0007 | -.0013 |
| 2.0000 | .0000 | .0006 | .0154 | .0025 | .0001 | .0007 | -.0011 |
| 3.0000 | .0000 | .0002 | .0119 | .0006 | -.0000 | .0003 | -.0004 |
| 4.0000 | .0000 | .0001 | .0090 | .0002 | -.0000 | .0001 | -.0002 |
| 5.0000 | .0000 | .0000 | .0068 | .0001 | -.0000 | .0000 | -.0001 |

TABLE A.3. INFLUENCE COEFFICIENTS FOR SIMPLY SUPPORTED COMPOSITE PLATES UNDER PARTIAL LOAD, EQU. (2.56).

$R = \dfrac{a}{b}$

$\xi' = .33 \quad (\xi' = \xi/a)$

$\eta' = .17 \quad (\eta' = \eta/b)$

$\nu = .30$

| F | KWR | KWS | KQX | KQY | KMX | KMY | KMXY |
|---|---|---|---|---|---|---|---|
| 1.0000 | .0002 | .0037 | .0166 | .0551 | .0028 | .0021 | -.0044 |
| 1.1000 | .0002 | .0035 | .0158 | .0465 | .0024 | .0021 | -.0034 |
| 1.2000 | .0002 | .0032 | .0149 | .0396 | .0021 | .0021 | -.0027 |
| 1.3000 | .0001 | .0029 | .0139 | .0340 | .0018 | .0020 | -.0022 |
| 1.4000 | .0001 | .0027 | .0130 | .0294 | .0016 | .0020 | -.0017 |
| 1.5000 | .0001 | .0025 | .0120 | .0256 | .0013 | .0019 | -.0014 |
| 1.6000 | .0001 | .0023 | .0111 | .0224 | .0012 | .0018 | -.0012 |
| 1.7000 | .0001 | .0021 | .0102 | .0197 | .0010 | .0017 | -.0010 |
| 1.8000 | .0001 | .0019 | .0094 | .0174 | .0009 | .0016 | -.0008 |
| 1.9000 | .0000 | .0018 | .0086 | .0155 | .0008 | .0015 | -.0007 |
| 2.0000 | .0000 | .0016 | .0073 | .0138 | .0007 | .0014 | -.0006 |
| 3.0000 | .0000 | .0008 | .0032 | .0052 | .0002 | .0008 | -.0001 |
| 4.0000 | .0000 | .0004 | .0013 | .0024 | .0001 | .0005 | -.0000 |
| 5.0000 | .0000 | .0002 | .0006 | .0013 | .0000 | .0003 | -.0000 |

TABLE A.4.

237

R = a/b = .50

$\xi' = .17 \quad (\xi' = \xi/a)$

$\eta' = .30 \quad (\eta' = \eta/b)$

$\nu =$

| R | KWB | KWS | KQX | KQY | KMX | KMY | KMXY |
|---|---|---|---|---|---|---|---|
| 1.0000 | .0003 | .0046 | .0114 | .1064 | .0037 | .0023 | -.0029 |
| 1.1500 | .0002 | .0044 | .0161 | .0937 | .0033 | .0023 | -.0022 |
| 1.2000 | .0002 | .0041 | .0089 | .0832 | .0030 | .0023 | -.0017 |
| 1.3000 | .0002 | .0038 | .0078 | .0743 | .0027 | .0023 | -.0013 |
| 1.4000 | .0001 | .0036 | .0068 | .0668 | .0024 | .0022 | -.0010 |
| 1.5000 | .0001 | .0034 | .0059 | .0603 | .0022 | .0022 | -.0008 |
| 1.6000 | .0001 | .0032 | .0051 | .0548 | .0020 | .0021 | -.0006 |
| 1.7000 | .0001 | .0030 | .0045 | .0499 | .0019 | .0020 | -.0005 |
| 1.8000 | .0001 | .0028 | .0039 | .0457 | .0017 | .0019 | -.0004 |
| 1.9000 | .0001 | .0026 | .0033 | .0420 | .0016 | .0018 | -.0003 |
| 2.0000 | .0001 | .0025 | .0029 | .0387 | .0015 | .0018 | -.0003 |
| 3.0000 | .0000 | .0015 | .0007 | .0197 | .0008 | .0012 | -.0000 |
| 4.0000 | .0000 | .0010 | .0002 | .0118 | .0005 | .0008 | -.0000 |
| 5.0000 | .0000 | .0007 | .0000 | .0078 | .0003 | .0006 | -.0000 |

TABLE A.5.

$$R = \frac{a}{b}$$

$\xi' = .17 \quad (\xi' = \xi/a)$

$\eta' = .33 \quad (\eta' = \eta/b)$

$\nu = .30$

| R | KWB | KWS | KQX | KQY | KMX | KMY | KMXY |
|---|---|---|---|---|---|---|---|
| 1.0000 | .0002 | .0037 | .0551 | .0166 | .0021 | .0028 | -.0044 |
| 1.1000 | .0002 | .0033 | .0536 | .0142 | .0016 | .0026 | -.0038 |
| 1.2000 | .0002 | .0029 | .0520 | .0122 | .0013 | .0025 | -.0033 |
| 1.3000 | .0001 | .0025 | .0503 | .0104 | .0010 | .0023 | -.0029 |
| 1.4000 | .0001 | .0022 | .0486 | .0086 | .0008 | .0021 | -.0025 |
| 1.5000 | .0001 | .0019 | .0469 | .0076 | .0006 | .0019 | -.0022 |
| 1.6000 | .0001 | .0017 | .0452 | .0066 | .0004 | .0018 | -.0020 |
| 1.7000 | .0001 | .0015 | .0435 | .0057 | .0003 | .0016 | -.0017 |
| 1.8000 | .0000 | .0013 | .0419 | .0049 | .0002 | .0015 | -.0015 |
| 1.9000 | .0000 | .0011 | .0404 | .0042 | .0002 | .0013 | -.0014 |
| 2.0000 | .0000 | .0010 | .0389 | .0037 | .0001 | .0012 | -.0012 |
| 3.0000 | .0000 | .0003 | .0267 | .0010 | -.0001 | .0004 | -.0005 |
| 4.0000 | .0000 | .0001 | .0188 | .0003 | -.0000 | .0002 | -.0002 |
| 5.0000 | .0000 | .0000 | .0136 | .0001 | -.0000 | .0001 | -.0001 |

TABLE A.6.

R = $\frac{a}{b}$

ξ' = .32 (ξ' = ξ/a)
η' = .33 (η' = η/b)
ν = .30

| F | KLA | KWS | KQX | KQY | KMX | KMY | KMXY |
|---|---|---|---|---|---|---|---|
| 1.0000 | .0005 | .0083 | .0370 | .0370 | .0054 | .0054 | -.0051 |
| 1.1000 | .0004 | .0075 | .0337 | .0321 | .0046 | .0052 | -.0042 |
| 1.2000 | .0003 | .0068 | .0307 | .0296 | .0039 | .0049 | -.0035 |
| 1.3000 | .0003 | .0052 | .0280 | .0266 | .0033 | .0047 | -.0029 |
| 1.4000 | .0002 | .0056 | .0254 | .0239 | .0029 | .0044 | -.0024 |
| 1.5000 | .0002 | .0050 | .0231 | .0215 | .0025 | .0041 | -.0020 |
| 1.6000 | .0001 | .0046 | .0210 | .0194 | .0021 | .0038 | -.0017 |
| 1.7000 | .0001 | .0042 | .0191 | .0175 | .0018 | .0036 | -.0014 |
| 1.8000 | .0001 | .0038 | .0174 | .0159 | .0016 | .0033 | -.0012 |
| 1.9000 | .0001 | .0035 | .0158 | .0144 | .0014 | .0031 | -.0010 |
| 2.0000 | .0001 | .0032 | .0144 | .0131 | .0012 | .0029 | -.0009 |
| 3.0000 | .0000 | .0014 | .0057 | .0058 | .0004 | .0015 | -.0002 |
| 4.0000 | .0000 | .0008 | .0024 | .0029 | .0001 | .0009 | -.0001 |
| 5.0000 | .0000 | .0004 | .0010 | .0017 | .0000 | .0005 | -.0000 |

TABLE A.7.

240

$$\frac{a}{b}$$

$R = $

$\xi' = .50 \quad (\xi' = \xi/a)$
$\eta' = .33 \quad (\eta' = \eta/b)$
$\nu = .30$

| R | KWB | KWS | KQX | KQY | KMX | KMY | KMXY |
|---|---|---|---|---|---|---|---|
| 1.0000 | .0005 | .0115 | .0215 | .0493 | .0083 | .0066 | -.0041 |
| 1.1000 | .0004 | .0106 | .0187 | .0454 | .0074 | .0064 | -.0032 |
| 1.2000 | .0004 | .0098 | .0162 | .0418 | .0065 | .0062 | -.0025 |
| 1.3000 | .0003 | .0091 | .0141 | .0386 | .0059 | .0059 | -.0020 |
| 1.4000 | .0003 | .0084 | .0121 | .0357 | .0053 | .0057 | -.0016 |
| 1.5000 | .0002 | .0078 | .0105 | .0331 | .0048 | .0054 | -.0013 |
| 1.6000 | .0002 | .0072 | .0091 | .0307 | .0043 | .0051 | -.0010 |
| 1.7000 | .0002 | .0067 | .0078 | .0286 | .0039 | .0048 | -.0008 |
| 1.8000 | .0001 | .0063 | .0068 | .0267 | .0036 | .0046 | -.0007 |
| 1.9000 | .0001 | .0059 | .0058 | .0249 | .0033 | .0043 | -.0005 |
| 2.0000 | .0001 | .0055 | .0050 | .0234 | .0030 | .0041 | -.0004 |
| 3.0000 | .0000 | .0032 | .0012 | .0133 | .0016 | .0026 | -.0001 |
| 4.0000 | .0000 | .0020 | .0003 | .0085 | .0009 | .0017 | -.0000 |
| 5.0000 | .0000 | .0014 | .0001 | .0059 | .0006 | .0013 | -.0000 |

TABLE A.8.

$R = \dfrac{a}{b}$

$\xi' = .17 \quad (\xi' = \xi/a)$
$\eta' = .50 \quad (\eta' = \eta/b)$
$\nu = .30$

| R | KWB | KWS | KQX | KQY | KMX | KMY | KMXY |
|---|---|---|---|---|---|---|---|
| 1.0000 | .0003 | .0046 | .1064 | .0114 | .0023 | .0037 | -.0029 |
| 1.1000 | .0002 | .0040 | .0994 | .0103 | .0018 | .0034 | -.0026 |
| 1.2000 | .0002 | .0035 | .0931 | .0093 | .0014 | .0031 | -.0023 |
| 1.3000 | .0002 | .0030 | .0874 | .0083 | .0011 | .0028 | -.0021 |
| 1.4000 | .0001 | .0026 | .0822 | .0074 | .0008 | .0026 | -.0018 |
| 1.5000 | .0001 | .0023 | .0774 | .0066 | .0006 | .0023 | -.0016 |
| 1.6000 | .0001 | .0020 | .0730 | .0058 | .0005 | .0021 | -.0015 |
| 1.7000 | .0001 | .0017 | .0690 | .0051 | .0003 | .0019 | -.0013 |
| 1.8000 | .0001 | .0015 | .0653 | .0045 | .0002 | .0017 | -.0012 |
| 1.9000 | .0000 | .0013 | .0619 | .0040 | .0002 | .0016 | -.0011 |
| 2.0000 | .0000 | .0012 | .0587 | .0035 | .0001 | .0014 | -.0010 |
| 3.0000 | .0000 | .0003 | .0366 | .0010 | -.0001 | .0005 | -.0004 |
| 4.0000 | .0000 | .0001 | .0244 | .0003 | -.0001 | .0002 | -.0002 |
| 5.0000 | .0000 | .0000 | .0171 | .0001 | -.0000 | .0001 | -.0001 |

TABLE A.9.

R $= \dfrac{a}{b}$

$\xi'$ = .33   ($\xi' = \xi/a$)
$\eta'$ = .50   ($\eta' = \eta/b$)
$\nu$ = .30

| R | KWB | KWS | KQX | KQY | KMX | KMY | KMXY |
|---|---|---|---|---|---|---|---|
| 1.0000 | .0005 | .0115 | .0493 | .0215 | .0066 | .0083 | -.0041 |
| 1.1000 | .0004 | .0102 | .0439 | .0201 | .0056 | .0077 | -.0035 |
| 1.2000 | .0004 | .0091 | .0392 | .0187 | .0047 | .0071 | -.0030 |
| 1.3000 | .0003 | .0081 | .0352 | .0173 | .0040 | .0066 | -.0025 |
| 1.4000 | .0002 | .0073 | .0316 | .0159 | .0034 | .0061 | -.0021 |
| 1.5000 | .0002 | .0065 | .0284 | .0147 | .0029 | .0056 | -.0018 |
| 1.6000 | .0002 | .0059 | .0256 | .0135 | .0025 | .0052 | -.0016 |
| 1.7000 | .0001 | .0053 | .0231 | .0124 | .0021 | .0048 | -.0013 |
| 1.8000 | .0001 | .0048 | .0209 | .0114 | .0019 | .0044 | -.0012 |
| 1.9000 | .0001 | .0044 | .0189 | .0105 | .0016 | .0041 | -.0010 |
| 2.0000 | .0001 | .0040 | .0172 | .0097 | .0014 | .0038 | -.0009 |
| 3.0000 | .0000 | .0018 | .0068 | .0047 | .0004 | .0019 | -.0002 |
| 4.0000 | .0000 | .0009 | .0029 | .0025 | .0001 | .0010 | -.0001 |
| 5.0000 | .0000 | .0005 | .0013 | .0015 | .0000 | .0006 | -.0000 |

TABLE A.10.

$$R = \frac{a}{b}$$

$\xi' = .50 \quad \{\xi' = \xi/a\}$
$\eta' = .50 \quad \{\eta' = \eta/b\}$
$\nu = .30$

| R | KWB | KWS | KQX | KQY | KMX | KMY | KMXY |
|---|---|---|---|---|---|---|---|
| 1.0000 | .0007 | .0182 | .0261 | .0261 | .0118 | .0118 | −.0037 |
| 1.1000 | .0005 | .0165 | .0224 | .0248 | .0104 | .0111 | −.0030 |
| 1.2000 | .0004 | .0150 | .0193 | .0234 | .0091 | .0104 | −.0024 |
| 1.3000 | .0004 | .0137 | .0166 | .0220 | .0081 | .0097 | −.0020 |
| 1.4000 | .0003 | .0125 | .0143 | .0207 | .0072 | .0091 | −.0016 |
| 1.5000 | .0003 | .0115 | .0123 | .0194 | .0065 | .0085 | −.0013 |
| 1.6000 | .0002 | .0106 | .0106 | .0182 | .0059 | .0079 | −.0011 |
| 1.7000 | .0002 | .0098 | .0092 | .0171 | .0053 | .0074 | −.0009 |
| 1.8000 | .0002 | .0091 | .0079 | .0161 | .0049 | .0070 | −.0007 |
| 1.9000 | .0001 | .0085 | .0068 | .0152 | .0044 | .0066 | −.0006 |
| 2.0000 | .0001 | .0079 | .0059 | .0144 | .0041 | .0062 | −.0005 |
| 3.0000 | .0000 | .0044 | .0014 | .0087 | .0020 | .0037 | −.0001 |
| 4.0000 | .0000 | .0028 | .0004 | .0058 | .0012 | .0024 | −.0000 |
| 5.0000 | .0000 | .0019 | .0001 | .0041 | .0008 | .0017 | −.0000 |

TABLE A.11.

R = $\frac{a}{b}$

$\xi' = .17$    $(\xi' = \xi/a)$
$\eta' = .17$    $(\eta' = \eta/b)$
$\nu = .30$

| R | KWR | KWS | KQX | KQY | KMX | KMY | KMXY |
|---|---|---|---|---|---|---|---|
| 1.0000 | .0023 | .0306 | .2556 | .2556 | .0199 | .0199 | -.1052 |
| 1.1000 | .0019 | .0275 | .2675 | .1997 | .0157 | .0201 | -.0864 |
| 1.2000 | .0015 | .0245 | .2762 | .1561 | .0122 | .0197 | -.0716 |
| 1.3000 | .0012 | .0216 | .2821 | .1265 | .0093 | .0189 | -.0597 |
| 1.4000 | .0010 | .0190 | .2855 | .1022 | .0070 | .0178 | -.0501 |
| 1.5000 | .0008 | .0167 | .2867 | .0833 | .0051 | .0166 | -.0424 |
| 1.6000 | .0007 | .0146 | .2859 | .0684 | .0036 | .0154 | -.0360 |
| 1.7000 | .0005 | .0128 | .2836 | .0566 | .0025 | .0141 | -.0308 |
| 1.8000 | .0004 | .0112 | .2799 | .0471 | .0016 | .0129 | -.0265 |
| 1.9000 | .0004 | .0097 | .2750 | .0394 | .0009 | .0118 | -.0229 |
| 2.0000 | .0003 | .0085 | .2691 | .0331 | .0004 | .0107 | -.0199 |
| 3.0000 | .0000 | .0022 | .1897 | .0068 | -.0009 | .0037 | -.0059 |
| 4.0000 | .0000 | .0006 | .1191 | .0016 | -.0005 | .0012 | -.0022 |
| 5.0000 | .0000 | .0002 | .0720 | .0005 | -.0002 | .0004 | -.0010 |

**TABLE A.12.** INFLUENCE COEFFICIENTS FOR SIMPLY SUPPORTED COMPOSITE PLATE UNDER CONCENTRATED LOAD, EQU. (2.64).

R = $\frac{a}{b}$

$\xi'$ = .33 ($\xi' = \xi/a$)
$\eta'$ = .17 ($\eta' = \eta/b$)
$\nu$ = .30

| R | KWB | KWS | KQX | KQY | KMX | KMY | KMXY |
|---|---|---|---|---|---|---|---|
| 1.0000 | .0041 | .0599 | .2712 | .9525 | .0457 | .0321 | -.0768 |
| 1.1000 | .0034 | .0557 | .2585 | .7061 | .0393 | .0331 | -.0592 |
| 1.2000 | .0028 | .0515 | .2434 | .5878 | .0337 | .0332 | -.0462 |
| 1.3000 | .0023 | .0473 | .2269 | .4924 | .0288 | .0327 | -.0365 |
| 1.4000 | .0019 | .0434 | .2101 | .4150 | .0246 | .0318 | -.0292 |
| 1.5000 | .0016 | .0397 | .1934 | .3518 | .0210 | .0306 | -.0235 |
| 1.6000 | .0013 | .0362 | .1773 | .2999 | .0179 | .0292 | -.0191 |
| 1.7000 | .0011 | .0331 | .1619 | .2570 | .0153 | .0278 | -.0157 |
| 1.8000 | .0009 | .0302 | .1475 | .2213 | .0130 | .0263 | -.0129 |
| 1.9000 | .0008 | .0276 | .1341 | .1915 | .0111 | .0248 | -.0107 |
| 2.0000 | .0007 | .0253 | .1217 | .1664 | .0094 | .0234 | -.0089 |
| 3.0000 | .0002 | .0107 | .0442 | .0495 | .0015 | .0124 | -.0018 |
| 4.0000 | .0000 | .0049 | .0158 | .0186 | -.0003 | .0066 | -.0005 |
| 5.0000 | .0000 | .0023 | .0058 | .0081 | -.0006 | .0036 | -.0001 |

TABLE A.13.

$$R = \frac{a}{b}$$

$\xi' = .50 \quad (\xi' = \xi/a)$

$\eta' = .17 \quad (\eta' = \eta/b)$

$\nu = .30$

| R | KWR | KWS | KQX | KQY | KMX | KMY | KMXY |
|---|---|---|---|---|---|---|---|
| 1.0000 | .0049 | .0759 | .1849 | 1.5505 | .0628 | .0358 | −.0488 |
| 1.1000 | .0041 | .0720 | .1641 | 1.4348 | .0564 | .0368 | −.0365 |
| 1.2000 | .0034 | .0681 | .1441 | 1.3343 | .0517 | .0368 | −.0276 |
| 1.3000 | .0028 | .0642 | .1257 | 1.2465 | .0472 | .0363 | −.0212 |
| 1.4000 | .0024 | .0606 | .1090 | 1.1692 | .0433 | .0354 | −.0164 |
| 1.5000 | .0020 | .0572 | .0942 | 1.1007 | .0400 | .0343 | −.0128 |
| 1.6000 | .0017 | .0541 | .0812 | 1.0396 | .0372 | .0331 | −.0101 |
| 1.7000 | .0014 | .0512 | .0699 | .9848 | .0347 | .0318 | −.0080 |
| 1.8000 | .0012 | .0485 | .0601 | .9355 | .0326 | .0305 | −.0063 |
| 1.9000 | .0011 | .0461 | .0516 | .8908 | .0307 | .0292 | −.0051 |
| 2.0000 | .0009 | .0439 | .0443 | .8502 | .0290 | .0280 | −.0041 |
| 3.0000 | .0003 | .0294 | .0046 | .5837 | .0190 | .0191 | −.0005 |
| 4.0000 | .0001 | .0220 | .0022 | .4440 | .0142 | .0143 | −.0001 |
| 5.0000 | .0001 | .0176 | .0006 | .3571 | .0114 | .0114 | −.0000 |

TABLE A.14.

$$R = \frac{a}{b}$$

$\xi' = .17 \quad (\xi' = \xi/a)$
$\eta' = .33 \quad (\eta' = \eta/b)$
$\nu = .30$

| R | KWB | KWS | KQX | KQY | KMX | KMY | KMXY |
|---|---|---|---|---|---|---|---|
| 1.0000 | .0041 | .0599 | .8535 | .2712 | .0321 | .0457 | -.0768 |
| 1.1000 | .0033 | .0524 | .8428 | .2308 | .0249 | .0431 | -.0674 |
| 1.2000 | .0027 | .0457 | .8245 | .1960 | .0191 | .0402 | -.0592 |
| 1.3000 | .0022 | .0397 | .8012 | .1663 | .0145 | .0372 | -.0521 |
| 1.4000 | .0018 | .0345 | .7744 | .1413 | .0108 | .0341 | -.0460 |
| 1.5000 | .0014 | .0300 | .7455 | .1202 | .0078 | .0312 | -.0407 |
| 1.6000 | .0012 | .0261 | .7155 | .1024 | .0055 | .0284 | -.0362 |
| 1.7000 | .0009 | .0226 | .6851 | .0875 | .0037 | .0257 | -.0322 |
| 1.8000 | .0008 | .0197 | .6547 | .0749 | .0023 | .0233 | -.0287 |
| 1.9000 | .0006 | .0171 | .6248 | .0642 | .0012 | .0213 | -.0256 |
| 2.0000 | .0005 | .0149 | .5955 | .0552 | .0004 | .0190 | -.0230 |
| 3.0000 | .0001 | .0038 | .3584 | .0136 | -.0015 | .0065 | -.0084 |
| 4.0000 | .0000 | .0010 | .2124 | .0039 | -.0008 | .0022 | -.0035 |
| 5.0000 | .0000 | .0003 | .1252 | .0013 | -.0003 | .0007 | -.0016 |

TABLE A.15.

248

R = $\frac{a}{b}$

ξ' = .33    (ξ' = ξ/a)
η' = .33    (η' = η/b)
ν = .30

| R | KWB | KWS | KQX | KQY | KMX | KMY | KMXY |
|---|---|---|---|---|---|---|---|
| 1.0000 | .0076 | .1324 | .5988 | .5988 | .0861 | .0861 | -.0874 |
| 1.1000 | .0062 | .1196 | .5450 | .5365 | .0721 | .0834 | -.0716 |
| 1.2000 | .0051 | .1078 | .4942 | .4797 | .0604 | .0797 | -.0588 |
| 1.3000 | .0042 | .0970 | .4471 | .4285 | .0506 | .0756 | -.0485 |
| 1.4000 | .0035 | .0874 | .4039 | .3827 | .0424 | .0711 | -.0401 |
| 1.5000 | .0029 | .0787 | .3645 | .3420 | .0356 | .0667 | -.0333 |
| 1.6000 | .0024 | .0709 | .3287 | .3059 | .0299 | .0623 | -.0278 |
| 1.7000 | .0020 | .0640 | .2964 | .2738 | .0251 | .0581 | -.0232 |
| 1.8000 | .0017 | .0578 | .2671 | .2454 | .0211 | .0540 | -.0195 |
| 1.9000 | .0014 | .0523 | .2407 | .2202 | .0178 | .0502 | -.0165 |
| 2.0000 | .0012 | .0474 | .2169 | .1979 | .0150 | .0467 | -.0139 |
| 3.0000 | .0003 | .0191 | .0768 | .0736 | .0020 | .0228 | -.0030 |
| 4.0000 | .0001 | .0085 | .0276 | .0310 | -.0007 | .0117 | -.0008 |
| 5.0000 | .0000 | .0040 | .0108 | .0143 | -.0011 | .0063 | -.0002 |

TABLE A.16.

R = $\frac{a}{b}$

ξ' = .50  (ξ' = ξ/a)
η' = .33  (η' = η/b)
ν = .30

| R | KWB | KWS | KQX | KQY | KMX | KMY | KMXY |
|---|---|---|---|---|---|---|---|
| 1.0000 | .0092 | .1894 | .3499 | 1.2182 | .1466 | .0997 | -.0678 |
| 1.1000 | .0076 | .1775 | .3029 | 1.1021 | .1328 | .0979 | -.0529 |
| 1.2000 | .0063 | .1663 | .2614 | 1.0028 | .1209 | .0952 | -.0415 |
| 1.3000 | .0052 | .1559 | .2250 | .9175 | .1108 | .0919 | -.0326 |
| 1.4000 | .0044 | .1464 | .1934 | .8437 | .1021 | .0883 | -.0258 |
| 1.5000 | .0037 | .1378 | .1660 | .7795 | .0945 | .0846 | -.0204 |
| 1.6000 | .0031 | .1299 | .1424 | .7234 | .0881 | .0809 | -.0163 |
| 1.7000 | .0026 | .1228 | .1221 | .6740 | .0824 | .0773 | -.0130 |
| 1.8000 | .0023 | .1163 | .1046 | .6303 | .0774 | .0738 | -.0104 |
| 1.9000 | .0019 | .1104 | .0897 | .5915 | .0731 | .0705 | -.0084 |
| 2.0000 | .0017 | .1051 | .0769 | .5569 | .0692 | .0674 | -.0068 |
| 3.0000 | .0005 | .0702 | .0166 | .3466 | .0456 | .0457 | -.0009 |
| 4.0000 | .0002 | .0526 | .0040 | .2496 | .0341 | .0343 | -.0001 |
| 5.0000 | .0001 | .0420 | .0016 | .1945 | .0272 | .0274 | -.0000 |

TABLE A.17.

$$R = \frac{a}{b} = .17 \quad \{\xi' = \xi/a\}$$

$$\xi' = .17 \quad \{\xi' = \xi/a\}$$
$$\eta' = .50 \quad \{\eta' = \eta/b\}$$
$$\nu = .30$$

| F | KWB | KWS | KQY | KQY | KMX | KMY | KMXY |
|---|---|---|---|---|---|---|---|
| 1.0000 | .0049 | .0759 | 1.5505 | .1849 | .0354 | .0628 | -.0448 |
| 1.1000 | .0039 | .0653 | 1.3815 | .1682 | .0290 | .0569 | -.0440 |
| 1.2000 | .0031 | .0564 | 1.2405 | .1512 | .0217 | .0516 | -.0397 |
| 1.3000 | .0025 | .0488 | 1.1210 | .1347 | .0167 | .0467 | -.0358 |
| 1.4000 | .0020 | .0422 | 1.0164 | .1192 | .0126 | .0423 | -.0323 |
| 1.5000 | .0017 | .0366 | .9293 | .1051 | .0094 | .0382 | -.0292 |
| 1.6000 | .0013 | .0318 | .8512 | .0924 | .0069 | .0345 | -.0265 |
| 1.7000 | .0011 | .0277 | .7821 | .0810 | .0049 | .0311 | -.0241 |
| 1.8000 | .0009 | .0241 | .7206 | .0709 | .0033 | .0281 | -.0219 |
| 1.9000 | .0007 | .0210 | .6654 | .0620 | .0021 | .0253 | -.0200 |
| 2.0000 | .0006 | .0184 | .6155 | .0543 | .0011 | .0228 | -.0182 |
| 3.0000 | .0001 | .0052 | .2976 | .0144 | -.0011 | .0060 | -.0077 |
| 4.0000 | .0000 | .0019 | .1419 | .0041 | -.0005 | .0029 | -.0035 |
| 5.0000 | .0000 | .0009 | .0574 | .0013 | -.0000 | .0012 | -.0017 |

TABLE A.18.

251

$$R = \frac{a}{b}$$

$$\xi' = .33 \quad \{\xi' = \xi/a\}$$
$$\eta' = .50 \quad \{\eta' = \eta/b\}$$
$$\nu = .30$$

| R | KWB | KWS | KQX | KQY | KMX | KMY | KMXY |
|---|---|---|---|---|---|---|---|
| 1.0000 | .0092 | .1894 | 1.2182 | .3499 | .0997 | .1466 | -.0678 |
| 1.1000 | .0075 | .1659 | 1.1087 | .3283 | .0826 | .1330 | -.0541 |
| 1.2000 | .0061 | .1459 | 1.0136 | .3054 | .0647 | .1210 | -.0496 |
| 1.3000 | .0050 | .1289 | .9303 | .2825 | .0573 | .1103 | -.0423 |
| 1.4000 | .0041 | .1143 | .8567 | .2603 | .0479 | .1007 | -.0360 |
| 1.5000 | .0034 | .1017 | .7914 | .2393 | .0402 | .0920 | -.0307 |
| 1.6000 | .0028 | .0907 | .7330 | .2197 | .0338 | .0842 | -.0262 |
| 1.7000 | .0023 | .0807 | .6807 | .2015 | .0285 | .0771 | -.0224 |
| 1.8000 | .0019 | .0729 | .6336 | .1848 | .0241 | .0706 | -.0192 |
| 1.9000 | .0016 | .0657 | .5912 | .1695 | .0204 | .0650 | -.0165 |
| 2.0000 | .0014 | .0593 | .5528 | .1554 | .0172 | .0598 | -.0141 |
| 3.0000 | .0003 | .0238 | .3149 | .0674 | .0030 | .0279 | -.0034 |
| 4.0000 | .0001 | .0110 | .2123 | .0310 | -.0001 | .0144 | -.0009 |
| 5.0000 | .0000 | .0056 | .1615 | .0149 | -.0006 | .0079 | -.0002 |

TABLE A.19.

R = $\dfrac{a}{b}$

ξ' = .50  (ξ' = ξ/a)
η' = .50  (η' = η/b)
ν = .30

| R | KWB | KWS | KQX | KQY | KMX | KMY | KMXY |
|---|---|---|---|---|---|---|---|
| 1.0000 | .0116 | .7057 | .6679 | .6679 | .4587 | .4587 | -.0610 |
| 1.1000 | .0095 | .6407 | .5977 | .6130 | .4104 | .4225 | -.0497 |
| 1.2000 | .0078 | .5853 | .5369 | .5639 | .3700 | .3909 | -.0403 |
| 1.3000 | .0065 | .5376 | .4843 | .5201 | .3359 | .3630 | -.0326 |
| 1.4000 | .0054 | .4962 | .4385 | .4811 | .3069 | .3362 | -.0264 |
| 1.5000 | .0045 | .4600 | .3987 | .4465 | .2819 | .3161 | -.0213 |
| 1.6000 | .0038 | .4281 | .3639 | .4157 | .2604 | .2962 | -.0173 |
| 1.7000 | .0033 | .3999 | .3335 | .3883 | .2416 | .2783 | -.0140 |
| 1.8000 | .0028 | .3747 | .3069 | .3638 | .2251 | .2621 | -.0113 |
| 1.9000 | .0024 | .3522 | .2836 | .3418 | .2105 | .2475 | -.0092 |
| 2.0000 | .0021 | .3320 | .2631 | .3221 | .1975 | .2341 | -.0075 |
| 3.0000 | .0006 | .2055 | .1508 | .2008 | .1189 | .1463 | -.0010 |
| 4.0000 | .0003 | .1448 | .1095 | .1445 | .0823 | .1058 | -.0002 |
| 5.0000 | .0001 | .1097 | .0884 | .1126 | .0616 | .0810 | -.0000 |

TABLE A.20.

$R = \dfrac{a}{b}$

$\xi' = .1C$

$\nu = .30$

$(\xi' = \xi/a)$

| F | KWB | KWS | KQX | KQY | KMX | KMY | KMXY |
|---|-----|-----|-----|-----|-----|-----|------|
| 1.0000 | .0009 | .0127 | .4210 | .0524 | .0067 | .0098 | .0288 |
| 1.1000 | .0007 | .0110 | .4118 | .0436 | .0051 | .0043 | .0247 |
| 1.2000 | .0006 | .0096 | .4026 | .0365 | .0038 | .0046 | .0214 |
| 1.3000 | .0005 | .0083 | .3934 | .0306 | .0028 | .0080 | .0187 |
| 1.4000 | .0004 | .0072 | .3843 | .0258 | .0020 | .0073 | .0165 |
| 1.5000 | .0003 | .0062 | .3753 | .0218 | .0014 | .0067 | .0146 |
| 1.6000 | .0003 | .0053 | .3664 | .0185 | .0009 | .0060 | .0130 |
| 1.7000 | .0002 | .0046 | .3576 | .0157 | .0005 | .0055 | .0116 |
| 1.8000 | .0002 | .0040 | .3489 | .0133 | .0002 | .0049 | .0104 |
| 1.9000 | .0001 | .0034 | .3404 | .0114 | .0000 | .0044 | .0094 |
| 2.0000 | .0001 | .0029 | .3320 | .0097 | -.0001 | .0039 | .0085 |
| 3.0000 | .0000 | .0007 | .2565 | .0021 | -.0004 | .0012 | .0037 |
| 4.0000 | .0000 | .0002 | .1965 | .0005 | -.0002 | .0004 | .0019 |
| 5.0000 | .0000 | .0000 | .1504 | .0001 | -.0001 | .0001 | .0011 |

TABLE A.21. INFLUENCE COEFFICIENTS FOR SIMPLY SUPPORTED COMPOSITE PLATE SUBJECTED TO STRIP LOAD, EQU. (2.72).

$$R = \frac{a}{b}$$

$\xi' = .70 \quad (\xi' = \xi/a)$

$\nu = .30$

| R | KWB | KWS | KQX | KQY | KMX | KMY | KMXY |
|---|---|---|---|---|---|---|---|
| 1.0000 | .0018 | .0265 | .3005 | .1161 | .0148 | .0196 | .0330 |
| 1.1000 | .0014 | .0233 | .2849 | .0975 | .0117 | .0186 | .0276 |
| 1.2000 | .0012 | .0204 | .2696 | .0823 | .0091 | .0174 | .0233 |
| 1.3000 | .0010 | .0179 | .2547 | .0698 | .0071 | .0162 | .0198 |
| 1.4000 | .0008 | .0157 | .2404 | .0595 | .0054 | .0150 | .0170 |
| 1.5000 | .0006 | .0137 | .2266 | .0508 | .0041 | .0138 | .0146 |
| 1.6000 | .0005 | .0120 | .2135 | .0436 | .0030 | .0126 | .0126 |
| 1.7000 | .0004 | .0105 | .2009 | .0375 | .0021 | .0115 | .0110 |
| 1.8000 | .0003 | .0092 | .1889 | .0324 | .0015 | .0105 | .0096 |
| 1.9000 | .0003 | .0080 | .1775 | .0280 | .0009 | .0095 | .0084 |
| 2.0000 | .0002 | .0070 | .1667 | .0243 | .0005 | .0086 | .0074 |
| 3.0000 | .0000 | .0020 | .0861 | .0064 | -.0006 | .0031 | .0024 |
| 4.0000 | .0000 | .0006 | .0417 | .0019 | -.0004 | .0011 | .0009 |
| 5.0000 | .0000 | .0002 | .0177 | .0006 | -.0002 | .0004 | .0004 |

TABLE A.22.

$$R = \frac{a}{b}$$

$\xi' = .30 \quad (\xi' = \xi/a)$
$\nu = .30$

| I | R | I | KWB | I | KWS | I | KQX | I | KQY | I | KMX | I | KMY | I | KMXY | I |
|---|---|---|---|---|---|---|---|---|---|---|---|---|---|---|---|---|
| I | 1.0000 | I | .0026 | I | .0424 | I | .2403 | I | .2095 | I | .0260 | I | .0291 | I | .0316 | I |
| I | 1.1000 | I | .0021 | I | .0379 | I | .2217 | I | .1782 | I | .0215 | I | .0278 | I | .0258 | I |
| I | 1.2000 | I | .0017 | I | .0339 | I | .2041 | I | .1526 | I | .0178 | I | .0263 | I | .0213 | I |
| I | 1.3000 | I | .0014 | I | .0302 | I | .1876 | I | .1314 | I | .0146 | I | .0247 | I | .0177 | I |
| I | 1.4000 | I | .0012 | I | .0270 | I | .1722 | I | .1138 | I | .0120 | I | .0231 | I | .0147 | I |
| I | 1.5000 | I | .0009 | I | .0241 | I | .1580 | I | .0990 | I | .0099 | I | .0215 | I | .0124 | I |
| I | 1.6000 | I | .0008 | I | .0216 | I | .1449 | I | .0864 | I | .0081 | I | .0199 | I | .0104 | I |
| I | 1.7000 | I | .0007 | I | .0193 | I | .1328 | I | .0758 | I | .0066 | I | .0185 | I | .0088 | I |
| I | 1.8000 | I | .0005 | I | .0173 | I | .1217 | I | .0666 | I | .0054 | I | .0171 | I | .0075 | I |
| I | 1.9000 | I | .0005 | I | .0155 | I | .1115 | I | .0588 | I | .0044 | I | .0158 | I | .0064 | I |
| I | 2.0000 | I | .0004 | I | .0139 | I | .1022 | I | .0520 | I | .0035 | I | .0146 | I | .0055 | I |
| I | 3.0000 | I | .0001 | I | .0051 | I | .0439 | I | .0173 | I | .0000 | I | .0066 | I | .0014 | I |
| I | 4.0000 | I | .0000 | I | .0020 | I | .0209 | I | .0066 | I | -.0005 | I | .0031 | I | .0004 | I |
| I | 5.0000 | I | .0000 | I | .0009 | I | .0120 | I | .0028 | I | -.0004 | I | .0015 | I | .0001 | I |

TABLE A.23.

$R = \dfrac{a}{b}$

$\xi' = .40 \quad (\xi' = \xi/a)$

$\nu = .30$

| R | KWR | KWS | KQX | KQY | KMX | KMY | KMXY |
|---|---|---|---|---|---|---|---|
| 1.0000 | .0031 | .0615 | .1683 | .3899 | .0419 | .0380 | .0278 |
| 1.1000 | .0026 | .0561 | .1498 | .3379 | .0364 | .0366 | .0222 |
| 1.2000 | .0021 | .0512 | .1328 | .2951 | .0317 | .0349 | .0179 |
| 1.3000 | .0018 | .0468 | .1175 | .2595 | .0277 | .0332 | .0145 |
| 1.4000 | .0015 | .0428 | .1036 | .2296 | .0243 | .0314 | .0118 |
| 1.5000 | .0012 | .0392 | .0912 | .2042 | .0214 | .0296 | .0097 |
| 1.6000 | .0010 | .0360 | .0802 | .1825 | .0189 | .0279 | .0079 |
| 1.7000 | .0009 | .0331 | .0703 | .1639 | .0168 | .0263 | .0065 |
| 1.8000 | .0007 | .0305 | .0616 | .1477 | .0149 | .0247 | .0054 |
| 1.9000 | .0006 | .0282 | .0539 | .1336 | .0133 | .0233 | .0045 |
| 2.0000 | .0005 | .0260 | .0470 | .1212 | .0119 | .0219 | .0037 |
| 3.0000 | .0001 | .0130 | .0163 | .0528 | .0043 | .0125 | .0007 |
| 4.0000 | .0001 | .0072 | -.0062 | .0269 | .0016 | .0077 | .0001 |
| 5.0000 | .0000 | .0042 | -.0032 | .0151 | .0004 | .0050 | .0000 |

TABLE A.24.

R $= \dfrac{a}{b}$

$\xi' = .50 \quad (\xi' = \xi/a)$

$\nu = .30$

| R | KWR | KWS | KQX | KQY | KMX | KMY | KMXY |
|---|---|---|---|---|---|---|---|
| 1.0000 | .0034 | .0834 | .1282 | 1.2201 | .0627 | .0457 | .0232 |
| 1.1000 | .0028 | .0777 | .1117 | 1.1016 | .0568 | .0443 | .0182 |
| 1.2000 | .0023 | .0725 | .0970 | 1.0017 | .0517 | .0426 | .0144 |
| 1.3000 | .0019 | .0678 | .0840 | .9166 | .0473 | .0408 | .0114 |
| 1.4000 | .0016 | .0635 | .0727 | .8432 | .0436 | .0390 | .0090 |
| 1.5000 | .0014 | .0597 | .0628 | .7795 | .0404 | .0372 | .0072 |
| 1.6000 | .0011 | .0562 | .0543 | .7237 | .0376 | .0354 | .0058 |
| 1.7000 | .0010 | .0530 | .0470 | .6744 | .0352 | .0338 | .0046 |
| 1.8000 | .0008 | .0502 | .0407 | .6307 | .0330 | .0322 | .0037 |
| 1.9000 | .0007 | .0476 | .0353 | .5917 | .0311 | .0307 | .0030 |
| 2.0000 | .0006 | .0452 | .0306 | .5567 | .0294 | .0293 | .0025 |
| 3.0000 | .0002 | .0299 | .0089 | .3398 | .0191 | .0198 | .0003 |
| 4.0000 | .0001 | .0222 | .0044 | .2365 | .0141 | .0148 | .0001 |
| 5.0000 | .0000 | .0176 | .0034 | .1774 | .0111 | .0118 | .0000 |

TABLE A.25.

258

$$R = \frac{a}{b}$$

$$\xi = .50 \quad (\xi' = \xi/a)$$

$$\nu = .30$$

| R | KWR | KWS | KQX | KQY | KMX | KMY | KMXY |
|---|---|---|---|---|---|---|---|
| 1.0000 | .0034 | .0834 | .1282 | 1.2201 | .0627 | .0457 | .0232 |
| 1.1000 | .0028 | .0777 | .1117 | 1.1016 | .0568 | .0443 | .0182 |
| 1.2000 | .0023 | .0725 | .0970 | 1.0017 | .0517 | .0426 | .0144 |
| 1.3000 | .0019 | .0678 | .0840 | .9166 | .0473 | .0408 | .0114 |
| 1.4000 | .0016 | .0635 | .0727 | .8432 | .0436 | .0390 | .0090 |
| 1.5000 | .0014 | .0597 | .0628 | .7795 | .0404 | .0372 | .0072 |
| 1.6000 | .0011 | .0562 | .0543 | .7237 | .0376 | .0354 | .0058 |
| 1.7000 | .0010 | .0530 | .0470 | .6744 | .0352 | .0338 | .0046 |
| 1.8000 | .0008 | .0502 | .0407 | .6307 | .0330 | .0322 | .0037 |
| 1.9000 | .0007 | .0476 | .0353 | .5917 | .0311 | .0307 | .0030 |
| 2.0000 | .0006 | .0452 | .0306 | .5567 | .0294 | .0293 | .0025 |
| 3.0000 | .0002 | .0299 | .0089 | .3398 | .0191 | .0198 | .0003 |
| 4.0000 | .0001 | .0222 | .0044 | .2365 | .0141 | .0148 | .0001 |
| 5.0000 | .0000 | .0176 | .0034 | .1774 | .0111 | .0118 | .0000 |

TABLE A.26.

259

# APPENDIX B

## Numerical Values for the Factors in Equs. (3.45) to (3.50) and (3.51) to (3.56)

| $R = \dfrac{a}{b}$ | $w \times \dfrac{D}{M_o\, b^2}$ | $\dfrac{M_x}{M_o}$ | $\dfrac{M_y}{M_o}$ |
|:---:|:---:|:---:|:---:|
| 1.0 | 0.037 | 0.394 | 0.256 |
| 1.2 | 0.052 | 0.420 | 0.393 |
| 1.4 | 0.067 | 0.424 | 0.515 |
| 1.6 | 0.079 | 0.415 | 0.619 |
| 1.8 | 0.087 | 0.402 | 0.703 |
| 2.0 | 0.096 | 0.387 | 0.770 |
| 3.0 | 0.117 | 0.331 | 0.939 |
| 4.0 | 0.122 | 0.309 | 0.985 |
| 5.0 | 0.125 | 0.302 | 0.996 |

TABLE B.1. DEFLECTIONS AND BENDING MOMENTS AT THE CENTER OF RECTANGULAR PLATES SUBJECTED TO UNIFORMLY DISTRIBUTED MOMENTS ALONG THE EDGES $y = \pm b/2$.

SPR......= 0.000
SPL......= 0.000
PL.......= 1.000
PR.......= 0.000
POIS.R...= .300
K........= .250

| RR | KWR | KMC | KQXR | KQYRN | KQYRF | KMXR | KMYR |
| RL | KWL |     | KQXL | KQYLN | KQYLF | KMXL | KMYL |
|---|---|---|---|---|---|---|---|
| 1.0000 | -.0001 |       | -.0024 | -.0553 | -.0011 | -.0006 | -.0004 |
| .2500  | -.0000 | -.006 | -.0001 | .2210  | .0000  | -.0000 | -.0000 |
| 1.2000 | -.0002 |       | -.0060 | -.0640 | -.0935 | -.0016 | -.0014 |
| .3000  | -.0000 | -.011 | -.0005 | .2549  | .0000  | -.0001 | -.0000 |
| 1.4000 | -.0004 |       | -.0111 | -.0751 | -.0081 | -.0031 | -.0034 |
| .3500  | -.0000 | -.018 | -.0017 | .2970  | .0000  | -.0004 | -.0001 |
| 1.6000 | -.0008 |       | -.0169 | -.0872 | -.0148 | -.0048 | -.0064 |
| .4000  | -.0000 | -.028 | -.0041 | .3404  | .0002  | -.0009 | -.0002 |
| 1.8000 | -.0013 |       | -.0228 | -.0992 | -.0232 | -.0068 | -.0103 |
| .4500  | -.0000 | -.038 | -.0078 | .3806  | -.0005 | -.0017 | .0002  |
| 2.0000 | -.0019 |       | -.0284 | -.1106 | -.0328 | -.0087 | -.0148 |
| .5000  | -.0001 | -.050 | -.0130 | .4151  | .0011  | -.0029 | -.0002 |
| 3.0000 | -.0049 |       | -.0441 | -.1529 | -.0822 | -.0169 | -.0392 |
| .7500  | -.0007 | -.100 | -.0496 | .4966  | .0124  | -.0120 | -.0034 |
| 4.0000 | -.0070 |       | -.0425 | -.1728 | -.1150 | -.0212 | -.0558 |
| 1.0000 | -.0018 | -.131 | -.0778 | .4860  | .0340  | -.0201 | -.0123 |
| 5.0000 | -.0080 |       | -.0340 | -.1799 | -.1309 | -.0231 | -.0639 |
| 1.2500 | -.0030 | -.145 | -.0864 | .4539  | .0547  | -.0240 | -.0219 |

TABLE B.2. NUMERICAL VALUES FOR FACTORS IN EQUS. (3.45) TO (3.50).

263

```
SPR......= .100
SPL......= .006
PL.......= 1.000
PR.......= 0.000
POIS.R...= .300
K........= .250
```

| RR / RL | KWR / KWL | KMC | KQXR / KQXL | KQYRN / KQYLN | KQYRF / KQYLF | KMXR / KMXL | KMYR / KMYL |
|---|---|---|---|---|---|---|---|
| 1.0000 | -.0000 |        | -.0008 | -.0046 | -.0004 | -.0000 | -.0003 |
|  .2500 | -.0000 | -.003  | -.0000 |  .0184 | -.0000 | -.0000 |  .0060 |
| 1.2000 | -.0001 |        | -.0025 | -.0106 | -.0015 | -.0002 | -.0010 |
|  .3000 | -.000C | -.006  | -.0052 |  .0419 | -.000C | -.0000 | -.0000 |
| 1.4000 | -.0002 |        | -.0054 | -.0192 | -.0039 | -.0006 | -.0025 |
|  .3500 | -.0000 | -.012  | -.0008 |  .0753 | -.0000 | -.0001 | -.0000 |
| 1.6000 | -.0005 |        | -.0092 | -.0297 | -.0080 | -.0014 | -.0048 |
|  .4000 | -.0000 | -.019  | -.0022 |  .1144 | -.0001 | -.0004 | -.0000 |
| 1.8000 | -.0008 |        | -.0136 | -.0411 | -.0138 | -.0023 | -.0078 |
|  .4500 | -.000C | -.027  | -.0047 |  .1549 | -.0003 | -.0009 | -.0000 |
| 2.0000 | -.0012 |        | -.0180 | -.0528 | -.0208 | -.0035 | -.0114 |
|  .5000 | -.0001 | -.036  | -.0383 |  .1938 | -.0007 | -.0016 | -.0001 |
| 3.0000 | -.0038 |        | -.0359 | -.1018 | -.0626 | -.0102 | -.0324 |
|  .7500 | -.0005 | -.081  | -.0378 |  .3196 | -.0095 | -.0084 | -.0033 |
| 4.0000 | -.0057 |        | -.0361 | -.1291 | -.0940 | -.0149 | -.0480 |
| 1.0000 | -.0015 | -.111  | -.0642 |  .3474 | -.0280 | -.0156 | -.0110 |
| 5.0000 | -.0068 |        | -.0308 | -.1412 | -.1105 | -.0172 | -.0561 |
| 1.2500 | -.0025 | -.125  | -.0747 |  .3380 |  .0469 | -.0196 | -.0197 |

TABLE B.3. NUMERICAL VALUES FOR FACTORS IN EQS. (3.45) TO (3.50).

SPR......= .200
SPL......= .013
PL.......= 1.000
PR.......= 0.000
POIS.R...= .300
K........= .250

| RR / RL | KWR / KWL | KMC | KQXR / KQXL | KQYRN / KQYLN | KQYRF / KQYLF | KMXR / KMXL | KMYR / KMYL |
|---|---|---|---|---|---|---|---|
| 1.0000 | -.0000 | -.002 | -.0005 | -.0027 | -.0002 | .0001 | -.0003 |
| .2500 | -.0000 | | -.0000 | .0108 | -.0000 | -.0000 | -.0000 |
| 1.2000 | -.0001 | -.005 | -.0016 | -.0066 | -.0009 | .0002 | -.0010 |
| .3000 | -.0000 | | -.0001 | .0262 | .0360 | -.0000 | -.0000 |
| 1.4000 | -.0001 | -.010 | -.0036 | -.0126 | -.0026 | -.0001 | -.0022 |
| .3500 | -.0000 | | -.0005 | .0494 | -.0000 | -.0001 | -.0000 |
| 1.6000 | -.0003 | -.016 | -.0063 | -.0203 | -.0055 | -.0001 | -.0041 |
| .4000 | -.0000 | | -.0015 | .0782 | .0061 | -.0002 | -.0000 |
| 1.8000 | -.0006 | -.023 | -.0097 | -.0291 | -.0098 | -.0005 | -.0067 |
| .4500 | -.0000 | | -.0033 | .1097 | .0302 | -.0011 | -.0001 |
| 2.0000 | -.0009 | -.031 | -.0132 | -.0385 | -.0152 | .0010 | -.0098 |
| .5000 | -.0000 | | -.0060 | .1414 | .0005 | -.0062 | -.0002 |
| 3.0000 | -.0030 | -.070 | -.0275 | -.0818 | -.0505 | -.0063 | -.0282 |
| .7500 | -.0004 | | -.0366 | .2565 | .0376 | -.0107 | -.0032 |
| 4.0000 | -.0049 | -.098 | -.0310 | -.1085 | -.0798 | -.0125 | -.0426 |
| 1.0000 | -.0013 | | -.0547 | .2906 | .0238 | -.0133 | -.0102 |
| 5.0000 | -.0059 | -.112 | -.0276 | -.1213 | -.0962 | -.0164 | -.0505 |
| 1.2500 | -.0022 | | -.0656 | .2870 | .0411 | | -.0180 |

TABLE B.4. NUMERICAL VALUES FOR FACTORS IN EQUS. (3.45) TO (3.50).

265

```
SPR......= .300
SPL......= .019
PL.......= 1.000
PR.......= 0.000
POIS.R...= .300
K........= .250
```

| RR / RL | KWR / KWL | KMC | KQXR / KQXL | KQYRN / KQYLN | KQYRF / KQYLF | KMXR / KMXL | KMYR / KMYL |
|---|---|---|---|---|---|---|---|
| 1.0000 | -.0000 | | -.0004 | -.0019 | -.0002 | .0001 | -.0003 |
| .2500 | -.0000 | -.002 | -.0000 | .0077 | -.0000 | -.0000 | -.0000 |
| 1.2000 | -.0000 | | -.0012 | -.0048 | -.0007 | -.0003 | -.0009 |
| .3000 | -.0000 | -.005 | -.0001 | .0191 | -.0000 | -.0000 | -.0000 |
| 1.4000 | -.0000 | | -.0024 | -.0094 | -.0019 | -.0005 | -.0021 |
| .3500 | -.0000 | -.009 | -.0074 | .0368 | -.0000 | -.0000 | -.0000 |
| 1.6000 | -.0002 | | -.0048 | -.0155 | -.0042 | -.0006 | -.0038 |
| .4000 | -.0000 | -.014 | -.0012 | .0595 | -.0000 | -.0001 | -.0001 |
| 1.8000 | -.0004 | | -.0075 | -.0226 | -.0076 | -.0005 | -.0061 |
| .4500 | -.0000 | -.020 | -.0026 | .0850 | -.0002 | -.0003 | -.0002 |
| 2.0000 | -.0007 | | -.0104 | -.0304 | -.0120 | -.0003 | -.0089 |
| .5000 | -.0000 | -.027 | -.0048 | .1114 | -.0004 | -.0007 | -.0003 |
| 3.0000 | -.0025 | | -.0231 | -.0685 | -.0424 | -.0034 | -.0254 |
| .7500 | -.0004 | -.063 | -.0257 | .2147 | -.0064 | -.0048 | -.0032 |
| 4.0000 | -.0042 | | -.0271 | -.0939 | -.0693 | -.0077 | -.0386 |
| 1.0000 | -.0011 | -.089 | -.0476 | .2509 | -.0267 | -.0102 | -.0095 |
| 5.0000 | -.0052 | | -.0248 | -.1069 | -.0853 | -.0104 | -.0461 |
| 1.2500 | -.0020 | -.102 | -.0584 | .2512 | .0365 | -.0139 | -.0167 |

TABLE B.5. NUMERICAL VALUES FOR FACTORS IN EQUS. (3.45) TO (3.50).

```
SPR......= .400
SPL......= .025
PL.......= 1.000
PR.......= 0.000
POIS.R...= .300
K........= .250
```

| RR RL | KHR KWL | KMC | KQXR KQXL | KQYRN KQYLN | KQYRF KQYLF | KMXR KMXL | KMYR KMYL |
|---|---|---|---|---|---|---|---|
| 1.0000 | -.0000 | | -.0003 | -.0015 | -.0001 | .0002 | -.0003 |
| .2500 | -.0000 | -.002 | -.0000 | .0060 | .0000 | .0000 | -.0060 |
| 1.2000 | -.0000 | | -.0009 | -.0038 | -.0005 | .0004 | -.0009 |
| .3000 | -.0000 | -.005 | -.0001 | .0150 | .0066 | -.0000 | -.0000 |
| 1.4000 | -.0001 | | -.0021 | -.0075 | -.0015 | -.0008 | -.0020 |
| .3500 | -.0000 | -.008 | -.0003 | .0293 | .0000 | -.0000 | -.0000 |
| 1.6000 | -.0002 | | -.0039 | -.0125 | -.0034 | -.0010 | -.0036 |
| .4000 | -.0000 | -.013 | -.0009 | .0480 | .0000 | -.0001 | -.0001 |
| 1.8000 | -.0004 | | -.0061 | -.0184 | -.0062 | -.0012 | -.0058 |
| .4500 | -.0000 | -.019 | -.0021 | .0694 | .0001 | -.0002 | -.0002 |
| 2.0000 | -.0006 | | -.0086 | -.0251 | -.0099 | -.0012 | -.0083 |
| .5000 | -.0000 | -.025 | -.0039 | .0920 | -.0003 | -.0004 | -.0004 |
| 3.0000 | -.0022 | | -.0199 | -.0589 | -.0365 | -.0015 | -.0234 |
| .7500 | -.0003 | -.058 | -.0221 | .1846 | -.0055 | -.0037 | -.0031 |
| 4.0000 | -.0037 | | -.0240 | -.0829 | -.0613 | -.0054 | -.0356 |
| 1.0000 | -.0010 | -.082 | -.0421 | .2210 | -.0183 | -.0085 | -.0090 |
| 5.0000 | -.0047 | | -.0225 | -.0957 | -.0766 | -.0681 | -.0427 |
| 1.2500 | -.0018 | -.095 | -.0527 | .2240 | .0329 | -.0119 | -.0156 |

TABLE B.6. NUMERICAL VALUES FOR FACTORS IN EQUS. (3.45) TO (3.50).

267

SPR.......= .500
SPL.......= .031
PL........= 1.000
PR........= 0.000
POIS.R....= .300
K.........= .250

| RR / RL | KWR / KWL | KMC | KQXR / KQXL | KQYRN / KQYLN | KQYRF / KQYLF | KMXR / KMXL | KMYR / KMYL |
|---|---|---|---|---|---|---|---|
| 1.0000 | -.0000 | | -.0002 | -.0012 | -.0001 | .0002 | -.0003 |
| .2500 | -.0000 | -.002 | -.0000 | -.0049 | -.0000 | .0000 | -.0000 |
| 1.2000 | -.0000 | | -.0008 | -.0031 | -.0004 | .0005 | -.0009 |
| .3000 | -.0001 | -.004 | -.0001 | .0124 | -.0000 | .0000 | -.0000 |
| 1.4000 | -.0000 | | -.0018 | -.0062 | -.0013 | .0009 | -.0019 |
| .3500 | -.0002 | -.008 | -.0003 | .0244 | -.0000 | -.0000 | -.0000 |
| 1.6000 | -.0003 | | -.0033 | -.0105 | -.0028 | .0013 | -.0035 |
| .4000 | -.0000 | -.012 | -.0008 | .0402 | -.0000 | -.0000 | -.0031 |
| 1.8000 | -.0003 | | -.0052 | -.0156 | -.0052 | .0016 | -.0055 |
| .4500 | -.0000 | -.018 | -.0018 | .0586 | -.0001 | -.0001 | -.0002 |
| 2.0000 | -.0005 | | -.0073 | -.0213 | -.0084 | .0018 | -.0079 |
| .5000 | -.0000 | -.024 | -.0034 | .0783 | -.0003 | -.0003 | -.0004 |
| 3.0000 | -.0019 | | -.0175 | -.0517 | -.0321 | -.0000 | -.0218 |
| .7500 | -.0003 | -.054 | -.0194 | .1620 | -.0049 | -.0029 | -.0031 |
| 4.0000 | -.0034 | | -.0216 | -.0742 | -.0550 | -.0035 | -.0332 |
| 1.0000 | -.0009 | -.076 | -.0378 | .1976 | .0164 | -.0071 | -.0086 |
| 5.0000 | -.0043 | | -.0206 | -.0867 | -.0696 | -.0063 | -.0398 |
| 1.2500 | -.0016 | -.088 | -.0480 | .2024 | .0299 | -.0103 | -.0147 |

TABLE B.7. NUMERICAL VALUES FOR FACTORS IN EQUS. (3.45) TO (3.50).

```
SPR......= 0.000
SPL......= 0.000
PL.......= 0.000
PR.......= 1.000
POIS.R...= .300
K........= .250
```

| RR / RL | KWR / KWL | KMC | KQXR / KQXL | KQYRN / KQYLN | KQYRF / KQYLF | KMXR / KMXL | KMYR / KMYL |
|---|---|---|---|---|---|---|---|
| 1.0000 | -.0013 |       | -.0574 | -.3819 | -.0251 | -.0149 | -.0091 |
| .2500  | -.0000 | -.098 | -.0022 | 1.5233 | -.0000 | -.0004 | .0002  |
| 1.2000 | -.0019 |       | -.0599 | -.3507 | -.0357 | -.0165 | -.0140 |
| .3000  | -.0000 | -.102 | -.0052 | 1.3924 | -.0001 | -.0011 | -.0004 |
| 1.4000 | -.0024 |       | -.0557 | -.3265 | -.0428 | -.0167 | -.0177 |
| .3500  | -.0000 | -.103 | -.0089 | 1.2878 | -.0002 | -.0019 | .0006  |
| 1.6000 | -.0026 |       | -.0483 | -.3096 | -.0469 | -.0163 | -.0200 |
| .4000  | -.0001 | -.102 | -.0127 | 1.2124 | -.0005 | -.0027 | .0006  |
| 1.8000 | -.0028 |       | -.0397 | -.2984 | -.0487 | -.0157 | -.0212 |
| .4500  | -.0001 | -.100 | -.0161 | 1.1602 | -.0010 | -.0035 | .0005  |
| 2.0000 | -.0028 |       | -.0314 | -.2910 | -.0491 | -.0151 | -.0217 |
| .5000  | -.0001 | -.099 | -.0187 | 1.1246 | -.0016 | -.0042 | .0003  |
| 3.0000 | -.0027 |       | -.0047 | -.2773 | -.0457 | -.0136 | -.0206 |
| .7500  | -.0003 | -.093 | -.0215 | 1.0564 | -.0056 | -.0055 | -.0015 |
| 4.0000 | -.0025 |       | .0024  | -.2738 | -.0433 | -.0133 | -.0194 |
| 1.0000 | -.0004 | -.091 | -.0161 | 1.0414 | -.0081 | -.0053 | -.0027 |
| 5.0000 | -.0025 |       | .0077  | -.2718 | -.0423 | -.0132 | -.0188 |
| 1.2500 | -.0005 | -.090 | -.0160 | 1.0346 | .0091  | -.0049 | -.0033 |

TABLE B.8. NUMERICAL VALUES FOR FACTORS IN EQUS. (3.45) TO (3.50).

TABLE B.9. NUMERICAL VALUES FOR FACTORS IN EQUS. (3.45) TO (3.50).

```
SPR...... = .100
SPL...... = .006
PL....... = 0.000
PR....... = 1.000
POIS.R... = .300
K........ = .250
```

| RR RL | KWR KWL | KMC | KQXR KQXL | KQYRN KQYLN | KQYRF KQYLF | KMXR KMXL | KMYR KMYL |
|---|---|---|---|---|---|---|---|
| 1.0000 | -.0004 | | -.0193 | -.0996 | -.0084 | -.0007 | -.0073 |
| .2500 | -.0000 | -.054 | -.0007 | -.3968 | -.0000 | -.0001 | .0000 |
| 1.2000 | -.0008 | | -.0253 | -.1070 | -.0148 | -.0021 | -.0105 |
| .3000 | -.0000 | -.061 | -.0021 | .4237 | -.0000 | -.0003 | .0001 |
| 1.4000 | -.0011 | | -.0281 | -.1075 | -.0207 | -.0033 | -.0132 |
| .3500 | -.0000 | -.064 | -.0043 | .4214 | .0001 | -.0007 | .0001 |
| 1.6000 | -.0014 | | -.0283 | -.1050 | -.0253 | -.0041 | -.0152 |
| .4000 | -.0000 | -.065 | -.0069 | .4059 | -.0303 | -.0012 | -.0061 |
| 1.8000 | -.0016 | | -.0267 | -.1017 | -.0284 | -.0046 | -.0164 |
| .4500 | -.0000 | -.065 | -.0095 | .3870 | .0006 | -.0018 | .0000 |
| 2.0000 | -.0017 | | -.0242 | -.0985 | -.0303 | -.0049 | -.0171 |
| .5000 | -.0001 | -.065 | -.0119 | .3693 | .0010 | -.0023 | -.0001 |
| 3.0000 | -.0018 | | -.0095 | -.0905 | -.0308 | -.0047 | -.0172 |
| .7500 | -.0002 | -.061 | -.0167 | .3225 | .0042 | -.0038 | -.0015 |
| 4.0000 | -.0017 | | -.0015 | -.0886 | -.0290 | -.0043 | -.0163 |
| 1.0000 | -.0004 | -.059 | -.0141 | .3120 | -.0067 | -.0039 | -.0026 |
| 5.0000 | -.0017 | | -.0010 | -.0880 | -.0281 | -.0042 | -.0159 |
| 1.2500 | -.0004 | -.059 | -.0058 | .3098 | .0077 | -.0036 | -.0032 |

SPR...... = .200
SPL...... = .013
PL....... = 0.000
PR....... = 1.000
POIS.R... = .300
K........ = .250

| RR<br>RL | KWR<br>KWL | KMC | KQXR<br>KQXL | KQYRN<br>KQYLN | KQYRF<br>KQYLF | KMXR<br>KMXL | KMYR<br>KMYL |
|---|---|---|---|---|---|---|---|
| 1.0000 | -.0003 |  | -.0116 | -.0595 | -.0050 | .0021 | -.0069 |
| .2500 | -.0000 | -.046 | -.0004 | .2373 | .0000 | -.0000 | -.0060 |
| 1.2000 | -.0005 |  | -.0159 | -.0669 | -.0093 | .0017 | -.0096 |
| .3000 | -.0000 | -.051 | -.0014 | .2650 | -.0000 | .0002 | -.0000 |
| 1.4000 | -.0008 |  | -.0186 | -.0698 | -.0137 | .0008 | -.0117 |
| .3500 | -.0000 | -.054 | -.0029 | .2733 | .0001 | -.0004 | -.0000 |
| 1.6000 | -.0010 |  | -.0195 | -.0701 | -.0174 | .0000 | -.0132 |
| .4000 | -.0000 | -.055 | -.0048 | .2705 | -.0002 | -.0007 | -.0001 |
| 1.8000 | -.0011 |  | -.0192 | -.0691 | -.0202 | -.0007 | -.0143 |
| .4500 | -.0000 | -.055 | -.0068 | .2625 | -.0004 | -.0011 | -.0002 |
| 2.0000 | -.0013 |  | -.0180 | -.0678 | -.0221 | -.0012 | -.0149 |
| .5000 | -.0001 | -.055 | -.0087 | .2529 | .0008 | -.0015 | -.0003 |
| 3.0000 | -.0014 |  | -.0049 | -.0628 | -.0243 | -.0018 | -.0152 |
| .7500 | -.0002 | -.053 | -.0135 | .2191 | .0034 | -.0028 | -.0014 |
| 4.0000 | -.0014 |  | -.0026 | -.0610 | -.0232 | -.0017 | -.0146 |
| 1.0000 | -.0003 | -.052 | -.0122 | .2084 | .0057 | -.0031 | -.0024 |
| 5.0000 | -.0013 |  | -.0000 | -.0604 | -.0225 | -.0015 | -.0143 |
| 1.2500 | -.0004 | -.051 | -.0090 | .2657 | .0067 | -.0029 | -.0030 |

TABLE B.10. NUMERICAL VALUES FOR FACTORS IN EQUS. (3.45) TO (3.50).

```
SPR...... = .300
SPL...... = .019
PL....... = 0.000
PR....... = 1.000
POIS.R... = .300
K........ = .250
```

| RR / RL | KWR / KWL | KMC | KQXR / KQXL | KQYRN / KQYLN | KQYRF / KQYLF | KMXR / KMXL | KMYR / KMYL |
|---|---|---|---|---|---|---|---|
| 1.0000 | -.0002 |  | -.0083 | -.0425 | -.0036 | .0033 | -.0068 |
| .2500 | -.0000 | -.043 | -.0003 | .1693 | .0000 | -.0000 | -.0003 |
| 1.2000 | -.0004 |  | -.0117 | -.0487 | -.0068 | .0034 | -.0092 |
| .3000 | -.0000 | -.047 | -.0010 | .1929 | .0000 | -.0001 | -.0001 |
| 1.4000 | -.0006 |  | -.0139 | -.0518 | -.0102 | .0029 | -.0110 |
| .3500 | -.0000 | -.049 | -.0021 | .2027 | -.0000 | .0002 | -.0031 |
| 1.6000 | -.0007 |  | -.0149 | -.0528 | -.0132 | .0022 | -.0122 |
| .4000 | -.0000 | -.050 | -.0036 | .2037 | .0001 | .0004 | -.0002 |
| 1.8000 | -.0009 |  | -.0150 | -.0527 | -.0156 | .0015 | -.0131 |
| .4500 | -.0000 | -.050 | -.0053 | .2000 | .0003 | .0007 | -.0003 |
| 2.0000 | -.0010 |  | -.0143 | -.0522 | -.0174 | .0009 | -.0136 |
| .5000 | -.0000 | -.050 | -.0069 | .1944 | -.0006 | .0010 | -.0005 |
| 3.0000 | -.0012 |  | -.0079 | -.0491 | -.0202 | .0002 | -.0139 |
| .7500 | -.0002 | -.048 | -.0114 | .1697 | .0029 | .0021 | -.0014 |
| 4.0000 | -.0012 |  | -.0029 | -.0477 | -.0196 | .0002 | -.0135 |
| 1.0000 | -.0003 | -.047 | -.0107 | .1601 | .0049 | .0025 | -.0023 |
| 5.0000 | -.0011 |  | -.0005 | -.0471 | -.0190 | .0001 | -.0132 |
| 1.2500 | -.0003 | -.047 | -.0082 | .1572 | .0060 | .0024 | -.0028 |

TABLE B.11. NUMERICAL VALUES FOR FACTORS IN EQUS. (3.45) TO (3.50).

```
SPR....... =   .400
SPL....... =   .025
PL........ = 0.000
PR........ = 1.000
POIS.R... =   .300
K........ =   .250
```

| RR / RL | KWR / KWL | KMC | KQXR / KQXL | KQYRN / KQYLN | KQYRF / KQYLF | KMXR / KMXL | KMYR / KMYL |
|---|---|---|---|---|---|---|---|
| 1.0000 | -.0001 |       | -.0064 | -.0330 | -.0028 |  .0040 | -.0067 |
|  .2500 | -.0000 | -.041 | -.0002 |  .1316 |  .0000 |  .0000 | -.0030 |
| 1.2000 | -.0003 |       | -.0092 | -.0383 | -.0054 | -.0044 | -.0089 |
|  .3000 | -.0000 | -.044 | -.0008 |  .1517 |  .0000 | -.0000 | -.0001 |
| 1.4000 | -.0004 |       | -.0111 | -.0412 | -.0081 |  .0041 | -.0105 |
|  .3500 | -.0000 | -.046 | -.0017 |  .1612 |  .0000 |  .0001 | -.0002 |
| 1.6000 | -.0006 |       | -.0120 | -.0424 | -.0107 |  .0035 | -.0116 |
|  .4000 | -.0000 | -.047 | -.0029 |  .1635 |  .0001 | -.0002 | -.0003 |
| 1.8000 | -.0007 |       | -.0122 | -.0427 | -.0128 |  .0029 | -.0123 |
|  .4500 | -.0000 | -.047 | -.0043 |  .1619 | -.0303 | -.0004 | -.0004 |
| 2.0000 | -.0008 |       | -.0119 | -.0425 | -.0144 |  .0023 | -.0127 |
|  .5000 | -.0010 | -.047 | -.0057 |  .1583 |  .0005 | -.0006 | -.0005 |
| 3.0000 | -.0010 |       | -.0070 | -.0406 | -.0173 | -.0019 | -.0130 |
|  .7500 | -.0001 | -.045 | -.0058 |  .1395 |  .0025 | -.0017 | -.0014 |
| 4.0000 | -.0010 |       | -.0029 | -.0396 | -.0171 |  .0008 | -.0126 |
| 1.0000 | -.0002 | -.044 | -.0056 |  .1312 |  .0044 | -.0020 | -.0022 |
| 5.0000 | -.0010 |       | -.0068 | -.0391 | -.0166 | -.0009 | -.0124 |
| 1.2500 | -.0003 | -.044 | -.0075 |  .1283 |  .0054 | -.0020 | -.0026 |

TABLE B.12. NUMERICAL VALUES FOR FACTORS IN EQUS. (3.45) TO (3.50).

```
SPR......= .500
SPL......= .031
PL.......= 0.000
PR.......= 1.000
POIS.R...= .300
K........= .250
```

| RR / RL | KWR / KWL | KMC | KQXR / KQXL | KQYRN / KQYLN | KQYRF / KQYLF | KMXR / KMXL | KMYR / KMYL |
|---|---|---|---|---|---|---|---|
| 1.0000 | -.0001 |  | -.0053 | -.0270 | -.0023 | .0044 | -.0066 |
| .2500 | -.0000 | -.039 | -.0062 | .1076 | .0000 | .0000 | -.0000 |
| 1.2000 | -.0002 |  | -.0076 | -.0316 | -.0344 | .0050 | -.0088 |
| .3000 | -.0000 | -.043 | -.0006 | .1250 | .0000 | .0000 | -.0001 |
| 1.4000 | -.0004 |  | -.0092 | -.0342 | -.0068 | .0049 | -.0103 |
| .3500 | -.0000 | -.044 | -.0014 | .1338 | .0000 | -.0000 | -.0002 |
| 1.6000 | -.0005 |  | -.0101 | -.0354 | -.0089 | .0044 | -.0112 |
| .4000 | -.0000 | -.044 | -.0025 | .1366 | .0061 | -.0001 | -.0003 |
| 1.8000 | -.0006 |  | -.0104 | -.0359 | -.0108 | .0038 | -.0118 |
| .4500 | -.0000 | -.044 | -.0036 | .1360 | .0062 | .0002 | -.0004 |
| 2.0000 | -.0007 |  | -.0161 | -.0359 | -.0122 | .0032 | -.0121 |
| .5000 | -.0000 | -.044 | -.0048 | .1336 | .0004 | -.0004 | -.0006 |
| 3.0000 | -.0009 |  | -.0062 | -.0348 | -.0151 | .0017 | -.0123 |
| .7500 | -.0061 | -.043 | -.0086 | .1189 | .0022 | -.0013 | -.0014 |
| 4.0000 | -.0009 |  | -.0028 | -.0340 | -.0151 | .0015 | -.0120 |
| 1.0000 | -.0002 | -.042 | -.0086 | .1116 | .0039 | .0017 | -.0021 |
| 5.0000 | -.0009 |  | -.0059 | -.0335 | -.0148 | .0016 | -.0118 |
| 1.2500 | -.0003 | -.042 | -.0069 | .1089 | .0049 | -.0017 | -.0025 |

TABLE B.13. NUMERICAL VALUES FOR FACTORS IN EQUS. (3.45) TO (3.50).

SPR.......= 0.000
SPL.......= 0.000
PL........= 1.000
PR........= 0.000
POIS.R....= .300
K.........= .500

| RR | KWR | ϑ | KMC | KQXR | KQYRN | KQYRF | KMXR | KMYR |
| RL | KWL |   |     | KQXL | KQYLN | KQYLF | KMXL | KMYL |
|---|---|---|---|---|---|---|---|---|
| 1.0000 | -.0006 |  |  | -.0272 | -.1939 | -.0118 | -.0070 | -.0043 |
| .5000 | -.0001 |  | -.045 | -.0118 | .3868 | .0010 | -.0026 | .0002 |
| 1.2000 | -.0013 |  |  | -.0403 | -.2115 | -.0235 | -.0107 | -.0093 |
| .6000 | -.0002 |  | -.063 | -.0233 | .4195 | .0034 | -.0054 | -.0004 |
| 1.4000 | -.0020 |  |  | -.0503 | -.2201 | -.0368 | -.0139 | -.0153 |
| .7000 | -.0004 |  | -.079 | -.0359 | .4325 | .0077 | -.0086 | -.0019 |
| 1.6000 | -.0028 |  |  | -.0564 | -.2233 | -.0497 | -.0163 | -.0215 |
| .8000 | -.0007 |  | -.091 | -.0477 | .4329 | .0137 | -.0117 | -.0041 |
| 1.8000 | -.0035 |  |  | -.0591 | -.2234 | -.0612 | -.0180 | -.0271 |
| .9000 | -.0011 |  | -.101 | -.0572 | .4261 | .0207 | -.0144 | -.0070 |
| 2.0000 | -.0041 |  |  | -.0591 | -.2219 | -.0707 | -.0192 | -.0320 |
| 1.0000 | -.0015 |  | -.109 | -.0641 | .4158 | .0280 | -.0166 | -.0102 |
| 3.0000 | -.0056 |  |  | -.0410 | -.2113 | -.0943 | -.0208 | -.0445 |
| 1.5000 | -.0032 |  | -.125 | -.0667 | .3658 | .0568 | -.0206 | -.0240 |
| 4.0000 | -.0060 |  |  | -.0218 | -.2054 | -.0992 | -.0208 | -.0473 |
| 2.0000 | -.0039 |  | -.127 | -.0476 | .3426 | .0682 | -.0202 | -.0304 |
| 5.0000 | -.0060 |  |  | -.0103 | -.2028 | -.0999 | -.0208 | -.0477 |
| 2.5000 | -.0042 |  | -.128 | -.0292 | .3342 | .0714 | -.0197 | -.0325 |

TABLE B.14. NUMERICAL VALUES FOR FACTORS IN EQUS. (3.45) TO (3.50).

```
SPR...... = .100
SPL...... = .025
PL....... = 1.000
PR....... = 0.000
POIS.R... = .300
K........ = .500
```

| RR RL | KWR KWL | KMC | KQXR KQXL | KQYRN KQYLN | KQYRF KQYLF | KMXR KMXL | KMYR KMYL |
|---|---|---|---|---|---|---|---|
| 1.0000 | -.0002 | | -.0091 | -.0460 | -.0039 | -.0003 | -.0034 |
| .5000 | -.0000 | -.025 | -.0039 | .0917 | .0063 | -.0004 | -.0004 |
| 1.2000 | -.0005 | | -.0167 | -.0677 | -.0098 | -.0014 | -.0069 |
| .6000 | -.0001 | -.038 | -.0097 | .1340 | .0014 | -.0013 | -.0011 |
| 1.4000 | -.0010 | | -.0245 | -.0859 | -.0178 | -.0029 | -.0113 |
| .7000 | -.0002 | -.051 | -.0174 | .1680 | .0037 | -.0028 | -.0023 |
| 1.6000 | -.0015 | | -.0309 | -.0996 | -.0270 | -.0045 | -.0160 |
| .8000 | -.0004 | -.062 | -.0259 | .1918 | .0074 | -.0046 | -.0040 |
| 1.8000 | -.0021 | | -.0355 | -.1093 | -.0361 | -.0061 | -.0205 |
| .9000 | -.0027 | -.071 | -.0340 | .2063 | .0122 | -.0064 | -.0062 |
| 2.0000 | -.0026 | | -.0382 | -.1158 | -.0445 | -.0075 | -.0245 |
| 1.0000 | -.0009 | -.079 | -.0407 | .2138 | .0177 | -.0082 | -.0087 |
| 3.0000 | -.0041 | | -.0341 | -.1252 | -.0689 | -.0111 | -.0361 |
| 1.5000 | -.0024 | -.096 | -.0515 | .2080 | .0423 | -.0130 | -.0200 |
| 4.0000 | -.0046 | | -.0214 | -.1250 | -.0751 | -.0118 | -.0391 |
| 2.0000 | -.0031 | -.099 | -.0411 | .1963 | .0537 | -.0135 | -.0258 |
| 5.0000 | -.0046 | | -.0116 | -.1246 | -.0762 | -.0119 | -.0396 |
| 2.5000 | -.0034 | -.100 | -.0275 | .1920 | .0573 | -.0133 | -.0279 |

TABLE B.15. NUMERICAL VALUES FOR FACTORS IN EQS. (3.45) TO (3.50).

```
SPR......= .200
SPL......= .050
PL.......= 1.000
PR.......= 0.000
POIS.R...= .300
K........= .500
```

| RR / RL | KWR / KWL | KMC | KQXR / KQXL | KQYRN / KQYLN | KQYRF / KQYLF | KMXR / KMXL | KMYR / KMYL |
|---|---|---|---|---|---|---|---|
| 1.0000 | -.0001 |        | -.0054 | -.0276 | -.0024 |  .0010 | -.0033 |
|  .5000 | -.0000 | -.021  | -.0024 |  .0550 |  .0002 | -.0000 | -.0005 |
| 1.2000 | -.0003 |        | -.0106 | -.0427 | -.0062 |  .0011 | -.0063 |
|  .6000 | -.0001 | -.032  | -.0061 |  .0844 |  .0009 | -.0003 | -.0012 |
| 1.4000 | -.0006 |        | -.0161 | -.0566 | -.0118 |  .0006 | -.0100 |
|  .7000 | -.0001 | -.042  | -.0115 |  .1106 |  .0025 | -.0009 | -.0024 |
| 1.6000 | -.0010 |        | -.0212 | -.0681 | -.0185 |  .0002 | -.0139 |
|  .8000 | -.0003 | -.052  | -.0178 |  .1312 |  .0051 | -.0019 | -.0040 |
| 1.8000 | -.0015 |        | -.0252 | -.0771 | -.0257 |  .0012 | -.0177 |
|  .9000 | -.0005 | -.060  | -.0241 |  .1456 |  .0087 |  .0031 | -.0059 |
| 2.0000 | -.0019 |        | -.0280 | -.0838 | -.0325 |  .0023 | -.0211 |
| 1.0000 | -.0007 | -.066  | -.0298 |  .1547 |  .0129 |  .0043 | -.0080 |
| 3.0000 | -.0033 |        | -.0278 | -.0965 | -.0548 |  .0063 | -.0312 |
| 1.5000 | -.0019 | -.082  | -.0414 |  .1592 |  .0338 |  .0087 | -.0176 |
| 4.0000 | -.0037 |        | -.0190 | -.0976 | -.0615 |  .0075 | -.0341 |
| 2.0000 | -.0026 | -.086  | -.0351 |  .1507 |  .0445 |  .0098 | -.0227 |
| 5.0000 | -.0038 |        | -.0111 | -.0975 | -.0630 |  .0078 | -.0346 |
| 2.5000 | -.0028 | -.086  | -.0247 |  .1467 |  .0482 |  .0098 | -.0246 |

TABLE B.16. NUMERICAL VALUES FOR FACTORS IN EQUS. (3.45) TO (3.50).

SPR.......= .300
SPL.......= .075
PL........= 1.000
PR........= 0.000
POIS.R...= .300
K.........= .500

| RR / RL | KWR / KWL | KMC | KQXR / KQXL | KQYRN / KQYLN | KQYRF / KQYLF | KMXR / KMXL | KMYR / KMYL |
|---|---|---|---|---|---|---|---|
| 1.0000 | -.0001 | | -.0039 | -.0197 | -.0017 | .0016 | -.0032 |
| .5000 | -.0001 | -.020 | -.0017 | -.0393 | -.0001 | .0002 | -.0005 |
| 1.2000 | -.0002 | | -.0077 | -.0312 | -.0045 | .0022 | -.0060 |
| .6000 | -.0000 | -.029 | -.0044 | -.0616 | -.0007 | .0002 | -.0013 |
| 1.4000 | -.0005 | | -.0120 | -.0422 | -.0088 | .0024 | -.0094 |
| .7000 | -.0001 | -.038 | -.0086 | -.0825 | -.0018 | -.0000 | -.0025 |
| 1.6000 | -.0008 | | -.0162 | -.0518 | -.0141 | .0021 | -.0128 |
| .8000 | -.0002 | -.046 | -.0136 | -.0997 | -.0039 | -.0005 | -.0040 |
| 1.8000 | -.0011 | | -.0196 | -.0597 | -.0199 | .0015 | -.0161 |
| .9000 | -.0004 | -.053 | -.0187 | -.1126 | -.0067 | .0013 | -.0057 |
| 2.0000 | -.0015 | | -.0221 | -.0658 | -.0256 | .0006 | -.0191 |
| 1.0000 | -.0005 | -.059 | -.0234 | -.1215 | -.0102 | -.0021 | -.0076 |
| 3.0000 | -.0027 | | -.0234 | -.0792 | -.0455 | -.0032 | -.0280 |
| 1.5000 | -.0016 | -.073 | -.0346 | -.1302 | -.0282 | -.0059 | -.0160 |
| 4.0000 | -.0032 | | -.0168 | -.0811 | -.0523 | -.0047 | -.0306 |
| 2.0000 | -.0022 | -.076 | -.0304 | -.1241 | -.0381 | -.0073 | -.0205 |
| 5.0000 | -.0033 | | -.0103 | -.0812 | -.0546 | -.0051 | -.0311 |
| 2.5000 | -.0025 | -.077 | -.0221 | -.1206 | .0417 | -.0075 | -.0222 |

TABLE B.17. NUMERICAL VALUES FOR FACTORS IN EQUS. (3.45) TO (3.50).

278

SPR......= .400
SPL......= .100
PL.......= 1.000
PR.......= 0.000
POIS.R...= .300
K........= .500

| RR / RL | KWR / KWL | KMC | KQXR / KQXL | KQYRN / KQYLN | KQYRF / KQYLF | KMXR / KMXL | KMYR / KMYL |
|---|---|---|---|---|---|---|---|
| 1.0000 | -.0001 |       | -.0030 | -.0153 | -.0013 | .0019 | -.0031 |
| .5000  | -.0003 | -.019 | -.0013 | .0305  | .0001  | .0001 | -.0006 |
| 1.2000 | -.0062 |       | -.0061 | -.0245 | -.0035 | .0028 | -.0059 |
| .6000  | -.0000 | -.028 | -.0035 | .0485  | .0005  | .0005 | -.0013 |
| 1.4000 | -.0004 |       | -.0096 | -.0336 | -.0070 | .0034 | -.0090 |
| .7000  | -.0001 | -.036 | -.0068 | .0658  | .0015  | .0005 | -.0025 |
| 1.6000 | -.0006 |       | -.0130 | -.0418 | -.0114 | .0035 | -.0121 |
| .8000  | -.0002 | -.043 | -.0109 | .0805  | .0031  | .0003 | -.0040 |
| 1.8000 | -.0009 |       | -.0160 | -.0487 | -.0162 | .0032 | -.0151 |
| .9000  | -.0003 | -.049 | -.0153 | .0919  | .0055  | -.0001 | -.0056 |
| 2.0000 | -.0012 |       | -.0182 | -.0542 | -.0211 | .0026 | -.0178 |
| 1.0000 | -.0005 | -.054 | -.0193 | -.1000 | .0084  | -.0007 | -.0073 |
| 3.0000 | -.0023 |       | -.0201 | -.0672 | -.0390 | .0010 | -.0257 |
| 1.5000 | -.0013 | -.067 | -.0296 | .1104  | .0241  | -.0039 | -.0149 |
| 4.0000 | -.0028 |       | -.0149 | -.0696 | -.0455 | -.0027 | -.0280 |
| 2.0000 | -.0019 | -.070 | -.0268 | .1060  | .0333  | -.0054 | -.0188 |
| 5.0000 | -.0029 |       | -.0095 | -.0699 | -.0473 | -.0032 | -.0285 |
| 2.5000 | -.0022 | -.070 | -.0199 | .1033  | .0368  | -.0058 | -.0203 |

TABLE B.18. NUMERICAL VALUES FOR FACTORS IN EQUS. (3.45) TO (3.50).

279

```
SPR......= .500
SPL......= .125
PL.......= 1.000
PR.......= 0.000
POIS.R...= .300
K........= .500
```

| RR / RL | KWR / KWL | KMC | KQXR / KQXL | KQYRN / KQYLN | KQYRF / KQYLF | KMXR / KMXL | KMYR / KMYL |
|---|---|---|---|---|---|---|---|
| 1.0000 | -.0001 |  | -.0025 | -.0125 | -.0011 | .0021 | -.0031 |
| .5000 | -.0000 |  | -.0011 | .0250 | .0001 | .0004 | -.0006 |
| 1.2000 | -.0000 | -.018 | -.0050 | -.0202 | -.0029 | .0033 | -.0058 |
| .6000 | -.0003 |  | -.0029 | .0400 | .0064 | .0007 | -.0014 |
| 1.4000 | -.0001 | -.026 | -.0080 | -.0281 | -.0058 | .0041 | -.0087 |
| .7000 | -.0005 |  | -.0109 | .0547 | .0012 | .0009 | -.0025 |
| 1.6000 | -.0001 | -.034 | -.0092 | -.0350 | -.0096 | .0044 | -.0117 |
| .8000 | -.0008 |  | -.0135 | .0674 | -.0026 | .0009 | -.0039 |
| 1.8000 | -.0002 | -.041 | -.0129 | -.0411 | -.0137 | .0043 | -.0145 |
| .9000 | -.0010 |  | -.0155 | .0776 | -.0046 | .0007 | -.0055 |
| 2.0000 | -.0004 | -.047 | -.0165 | -.0461 | -.0180 | .0039 | -.0169 |
| 1.0000 | -.0020 |  | -.0177 | .0853 | -.0072 | .0003 | -.0071 |
| 3.0000 | -.0012 | -.051 | -.0260 | -.0585 | -.0341 | .0307 | -.0239 |
| 1.5000 | -.0025 |  | -.0134 | .0959 | .0211 | -.0024 | -.0140 |
| 4.0000 | -.0017 | -.062 | -.0239 | -.0611 | -.0403 | -.0012 | -.0260 |
| 2.0000 | -.0026 |  | -.0087 | .0927 | .0295 | -.0040 | -.0175 |
| 5.0000 | -.0019 | -.065 | -.0180 | -.0615 | -.0421 | -.0018 | -.0264 |
| 2.5000 |  |  | -.0160 | .0902 | .0329 | -.0045 | -.0188 |

TABLE B.19. NUMERICAL VALUES FOR FACTORS IN EQUS. (3.45) TO (3.50).

SPR..... = 0.000
SPL..... = 0.000
PL...... = 0.000
PR...... = 1.000
POIS.R.. = .300
K....... = .500

| RR / RL | KWR / KWL | KMC | KQXR / KQXL | KQYRN / KQYLN | KQYRF / KQYLF | KMXR / KMXL | KMYR / KMYL |
|---|---|---|---|---|---|---|---|
| 1.0000 | -.0013 |       | -.0574 | -.3819 | -.0251 | -.0149 | -.0091 |
|  .5000 | -.0002 | -.098 | -.0250 |  .7617 |  .0022 | -.0056 |  .0004 |
| 1.2000 | -.0019 |       | -.0599 | -.3508 | -.0357 | -.0165 | -.0140 |
|  .6000 | -.0003 | -.102 | -.0352 |  .6964 |  .0052 | -.0081 | -.0006 |
| 1.4000 | -.0024 |       | -.0558 | -.3267 | -.0429 | -.0168 | -.0177 |
|  .7000 | -.0005 | -.103 | -.0415 |  .6444 |  .0089 | -.0100 |  .0021 |
| 1.6000 | -.0026 |       | -.0484 | -.3101 | -.0470 | -.0163 | -.0251 |
|  .8000 | -.0007 | -.102 | -.0400 |  .6074 |  .0128 | -.0110 | -.0038 |
| 1.8000 | -.0028 |       | -.0439 | -.2992 | -.0489 | -.0158 | -.0214 |
|  .9000 | -.0009 | -.101 | -.0319 |  .5822 |  .0162 | -.0115 | -.0054 |
| 2.0000 | -.0029 |       | -.0418 | -.2923 | -.0496 | -.0152 | -.0220 |
| 1.0000 | -.0010 | -.099 |        |  .5655 |  .0190 | -.0116 | -.0068 |
| 3.0000 | -.0028 |       | -.0060 | -.2811 | -.0481 | -.0141 | -.0217 |
| 1.5000 | -.0014 | -.096 | -.0227 |  .5371 |  .0251 | -.0105 | -.0102 |
| 4.0000 | -.0028 |       | -.0008 | -.2790 | -.0472 | -.0140 | -.0213 |
| 2.0000 | -.0015 | -.095 | -.0090 |  .5322 |  .0258 | -.0099 | -.0109 |
| 5.0000 | -.0028 |       | -.0013 | -.2777 | -.0470 | -.0140 | -.0212 |
| 2.5000 | -.0015 | -.095 | -.0029 |  .5295 |  .0258 | -.0098 | -.0110 |

TABLE B.20. NUMERICAL VALUES FOR FACTORS IN EQUS. (3.45) TO (3.50).

```
SPR.......= .100
SPL.......= .025
PL........= 0.000
PR........= 1.000
POIS.R....= .300
K.........= .500
```

| RR RL | KHR KWL | KMC | KQXR KQXL | KQYRN KQYLN | KQYRF KQYLF | KMXR KMXL | KMYR KMYL |
|---|---|---|---|---|---|---|---|
| 1.0000 | -.0004 | | -.0193 | -.0996 | -.0084 | -.0007 | -.0073 |
| .5000 | -.0001 | -.054 | -.0083 | .1984 | -.0007 | -.0009 | -.0008 |
| 1.2000 | -.0008 | | -.0253 | -.1070 | -.0148 | -.0021 | -.0105 |
| .6000 | -.0001 | -.061 | -.0146 | .2119 | .0021 | -.0020 | -.0016 |
| 1.4000 | -.0011 | | -.0281 | -.1076 | -.0207 | -.0033 | -.0132 |
| .7000 | -.0002 | -.064 | -.0202 | .2108 | -.0043 | .0032 | .0026 |
| 1.6000 | -.0014 | | -.0283 | -.1051 | -.0253 | -.0041 | -.0152 |
| .8000 | -.0004 | -.065 | -.0242 | .2033 | .0370 | -.0043 | -.0038 |
| 1.8000 | -.0016 | | -.0268 | -.1020 | -.0285 | -.0047 | -.0165 |
| .9000 | -.0005 | -.065 | -.0265 | .1943 | -.0096 | -.0051 | -.0049 |
| 2.0000 | -.0018 | | -.0243 | -.0990 | -.0305 | -.0049 | -.0172 |
| 1.0000 | -.0006 | -.065 | -.0272 | .1860 | .0120 | -.0056 | -.0059 |
| 3.0000 | -.0019 | | -.0102 | -.0925 | -.0321 | -.0049 | -.0178 |
| 1.5000 | -.0010 | -.063 | -.0201 | .1666 | .0185 | -.0059 | -.0089 |
| 4.0000 | -.0019 | | -.0025 | -.0918 | -.0314 | -.0047 | -.0175 |
| 2.0000 | -.0011 | -.062 | -.0105 | .1640 | .0197 | -.0055 | -.0096 |
| 5.0000 | -.0019 | | -.0000 | -.0918 | -.0312 | -.0047 | -.0174 |
| 2.5000 | -.0012 | -.062 | -.0044 | .1638 | .0198 | -.0053 | -.0098 |

TABLE B.21. NUMERICAL VALUES FOR FACTORS IN EQUS. (3.45) TO (3.50).

```
SPR......= .200
SPL......= .050
PL.......= 0.000
PR.......= 1.000
POIS.R...= .300
K........= .500
```

| RR / RL | KWR / KWL | KMC | KQXR / KQXL | KQYRN / KQYLN | KQYRF / KQYLF | KMXR / KMXL | KMYR / KMYL |
|---|---|---|---|---|---|---|---|
| 1.0000 | -.0003 | -.046 | -.0116 | -.0595 | -.0050 | .0021 | -.0069 |
| .5000 | -.0000 |  | -.0050 | -.1186 | .0004 | -.0000 | -.0010 |
| 1.2000 | -.0005 | -.051 | -.0159 | -.0669 | -.0093 | .0017 | -.0096 |
| .6000 | -.0001 |  | -.0092 | -.1325 | .0014 | -.0004 | -.0019 |
| 1.4000 | -.0008 | -.054 | -.0166 | -.0698 | -.0137 | -.0008 | -.0117 |
| .7000 | -.0002 |  | -.0133 | -.1367 | -.0029 | .0011 | -.0028 |
| 1.6000 | -.0010 | -.055 | -.0196 | -.0701 | -.0174 | .0000 | -.0133 |
| .8000 | -.0003 |  | -.0166 | -.1354 | .0048 | -.0018 | -.0037 |
| 1.8000 | -.0012 | -.055 | -.0152 | -.0692 | -.0242 | -.0007 | -.0143 |
| .9000 | -.0004 |  | -.0168 | -.1317 | .0068 | -.0024 | -.0046 |
| 2.0000 | -.0013 | -.055 | -.0181 | -.0680 | -.0222 | -.0012 | -.0149 |
| 1.0000 | -.0015 |  | -.0200 | -.1273 | .0088 | -.0029 | -.0055 |
| 3.0000 | -.0015 | -.054 | -.0093 | -.0640 | -.0251 | -.0019 | -.0156 |
| 1.5000 | -.0008 |  | -.0166 | -.1132 | .0147 | -.0038 | -.0079 |
| 4.0000 | -.0015 | -.054 | -.0033 | -.0631 | -.0248 | -.0019 | -.0155 |
| 2.0000 | -.0009 |  | -.0097 | -.1101 | .0162 | -.0036 | -.0086 |
| 5.0000 | -.0015 | -.053 | -.0037 | -.0630 | -.0247 | -.0019 | -.0154 |
| 2.5000 | -.0010 |  | -.0047 | -.1097 | .0164 | -.0035 | -.0088 |

TABLE B.22. NUMERICAL VALUES FOR FACTORS IN EQUS. (3.45) TO (3.50).

SPR.......= .300
SPL.......= .075
PL........= 0.000
PR........= 1.000
POIS.R....= .300
K.........= .500

| RR / RL | KWR / KWL | KMC | KQXR / KQXL | KQYRN / KQYLN | KQYRF / KQYLF | KMXR / KMXL | KMYR / KMYL |
|---|---|---|---|---|---|---|---|
| 1.0000 | -.0002 | -.043 | -.0083 | -.0425 | -.0036 | .0033 | -.0068 |
| .5000 | -.0000 | | -.0036 | -.0846 | .0003 | .0004 | -.0011 |
| 1.2000 | -.0004 | -.047 | -.0117 | -.0487 | -.0068 | .0034 | -.0092 |
| .6000 | -.0001 | | -.0067 | .0965 | .0010 | .0003 | -.0020 |
| 1.4000 | -.0006 | -.049 | -.0139 | -.0518 | -.0102 | .0029 | -.0110 |
| .7000 | -.0001 | | -.0100 | .1014 | .0021 | .0000 | -.0029 |
| 1.6000 | -.0007 | -.050 | -.0149 | -.0528 | -.0132 | .0022 | -.0122 |
| .8000 | -.0002 | | -.0127 | .1020 | .0036 | -.0005 | -.0037 |
| 1.8000 | -.0009 | -.050 | -.0150 | -.0528 | -.0157 | .0015 | -.0131 |
| .9000 | -.0003 | | -.0146 | .1003 | .0053 | .0010 | -.0045 |
| 2.0000 | -.0010 | -.050 | -.0144 | -.0523 | -.0175 | .0009 | -.0136 |
| 1.0000 | -.0004 | | -.0158 | .0977 | .0069 | .0014 | -.0052 |
| 3.0000 | -.0012 | -.049 | -.0081 | -.0499 | -.0207 | .0002 | -.0142 |
| 1.5000 | -.0007 | | -.0140 | .0875 | .0122 | -.0024 | -.0072 |
| 4.0000 | -.0012 | -.049 | -.0033 | -.0492 | -.0207 | .0003 | -.0141 |
| 2.0000 | -.0038 | | -.0087 | .0846 | .0137 | .0025 | -.0079 |
| 5.0000 | -.0012 | -.049 | -.0010 | -.0491 | -.0266 | -.0003 | -.0140 |
| 2.5000 | -.0008 | | -.0046 | .0841 | .0140 | -.0024 | -.0080 |

TABLE B.23. NUMERICAL VALUES FOR FACTORS IN EQUS. (3.45) TO (3.50).

284

SPR......= .400
SPL......= .100
PL.......= 0.000
PR.......= 1.000
POIS.R...= .300
K........= .500

| RR / RL | KWR / KWL | KMC | KQXR / KQXL | KQYRN / KQYLN | KQYRF / KQYLF | KMXR / KMXL | KMYR / KMYL |
|---|---|---|---|---|---|---|---|
| 1.0000 | -.0001 |      | -.0064 | -.0330 | -.0028 |  .0040 | -.0067 |
|  .5000 | -.0000 | -.041 | -.0028 |  .0658 |  .0002 |  .0006 | -.0012 |
| 1.2000 | -.0303 |      | -.0092 | -.0383 | -.0054 |  .0044 | -.0089 |
|  .6000 | -.0000 | -.044 | -.0053 |  .0759 | -.0008 |  .0007 | -.0020 |
| 1.4000 | -.0004 |      | -.0111 | -.0412 | -.0081 |  .0041 | -.0105 |
|  .7000 | -.0001 | -.046 | -.0079 |  .0806 |  .0017 |  .0006 | -.0029 |
| 1.6000 | -.0006 |      | -.0121 | -.0424 | -.0107 |  .0035 | -.0116 |
|  .8000 | -.0002 | -.047 | -.0102 |  .0818 |  .0029 |  .0003 | -.0037 |
| 1.8000 | -.0007 |      | -.0123 | -.0427 | -.0128 |  .0029 | -.0123 |
|  .9000 | -.0002 | -.047 | -.0119 |  .0811 | -.0043 | -.0000 | -.0044 |
| 2.0000 | -.0008 |      | -.0130 | -.0426 | -.0144 |  .0023 | -.0127 |
| 1.0000 | -.0003 | -.047 | -.0072 |  .0795 | -.0057 | -.0004 | -.0050 |
| 3.0000 | -.0011 |      | -.0121 | -.0412 | -.0176 |  .0009 | -.0132 |
| 1.5000 | -.0006 | -.046 | -.0032 |  .0718 | -.0105 | -.0015 | -.0067 |
| 4.0000 | -.0011 |      | -.0078 | -.0406 | -.0179 | -.0007 | -.0131 |
| 2.0000 | -.0007 | -.045 | -.0012 |  .0693 |  .0120 | -.0017 | -.0073 |
| 5.0000 | -.0011 |      |        | -.0405 | -.0178 | -.0007 | -.0131 |
| 2.5000 | -.0007 | -.045 | -.0043 |  .0688 |  .0123 | -.0016 | -.0075 |

TABLE B.24. NUMERICAL VALUES FOR FACTORS IN EQUS. (3.45) TO (3.50).

| RR / RL | KWR / KWL | KMC | KQXR / KQXL | KQYRN / KQYLN | KQYRF / KQYLF | KMXR / KMXL | KMYR / KMYL |
|---|---|---|---|---|---|---|---|
| 1.0000 | -.0001 |       | -.0053 | -.0270 | -.0023 | .0044 | -.0066 |
| .5000  | -.0000 | -.039 | -.0023 | .0538  | .0302  | .0007 | -.0012 |
| 1.2000 | -.0002 |       | -.0076 | -.0316 | -.0044 | .0050 | -.0088 |
| .6000  | -.0000 | -.043 | -.0044 | .0625  | .0006  | .0010 | -.0021 |
| 1.4000 | -.0004 |       | -.0092 | -.0342 | -.0068 | .0049 | -.0103 |
| .7000  | -.0001 | -.044 | -.0066 | .0669  | .0014  | .0010 | -.0029 |
| 1.6000 | -.0005 |       | -.0101 | -.0354 | -.0089 | .0044 | -.0112 |
| .8000  | -.0001 | -.044 | -.0086 | .0684  | .0025  | .0009 | -.0037 |
| 1.8000 | -.0006 |       | -.0104 | -.0359 | -.0108 | .0038 | -.0118 |
| .9000  | -.0002 | -.044 | -.0101 | .0682  | .0036  | .0006 | -.0044 |
| 2.0000 | -.0007 |       | -.0101 | .0360  | -.0123 | .0032 | -.0121 |
| 1.0000 | -.0003 | -.044 | -.0111 | .0671  | .0049  | .0003 | -.0049 |
| 3.0000 | -.0009 |       | -.0064 | -.0351 | -.0153 | .0017 | -.0124 |
| 1.5000 | -.0005 | -.043 | -.0106 | .0611  | .0092  | -.0008 | -.0064 |
| 4.0000 | -.0009 |       | -.0030 | -.0347 | -.0157 | .0015 | -.0123 |
| 2.0000 | -.0006 | -.043 | -.0071 | .0589  | .0106  | -.0011 | -.0069 |
| 5.0000 | -.0009 |       | -.0012 | -.0346 | -.0157 | .0015 | -.0123 |
| 2.5000 | -.0006 | -.043 | -.0040 | .0583  | .0109  | -.0011 | -.0070 |

TABLE B.25. NUMERICAL VALUES FOR FACTORS IN EQUS. (3.45) TO (3.50).

```
SPR......= 0.000
SPL......= 0.000
PL.......= 1.000
PR.......= 0.000
POIS.R...= .300
K........= .750
```

| RR / RL | KWR / KWL | KMC | KQXR / KQXL | KQYRN / KQYLN | KQYRF / KQYLF | KMXR / KMXL | KMYR / KMYL |
|---|---|---|---|---|---|---|---|
| 1.0000 | -.0012 |       | -.0496 | -.3157 | -.0216 | -.0127 | -.0078 |
| .7500  | -.0006 | -.082 | -.0403 | .4198  | .0101  | -.0097 | -.0028 |
| 1.2000 | -.0019 |       | -.0596 | -.3062 | -.0350 | -.0161 | -.0138 |
| .9000  | -.0010 | -.095 | -.0537 | .4055  | .0194  | -.0135 | -.0066 |
| 1.4000 | -.0026 |       | -.0629 | -.2930 | -.0468 | -.0179 | -.0195 |
| 1.0500 | -.0016 | -.104 | -.0614 | .3856  | .0292  | -.0161 | -.0109 |
| 1.6000 | -.0031 |       | -.0613 | -.2809 | -.0559 | -.0187 | -.0241 |
| 1.2000 | -.0021 | -.108 | -.0639 | .3669  | .0380  | -.0176 | -.0150 |
| 1.8000 | -.0036 |       | -.0568 | -.2711 | -.0624 | -.0190 | -.0275 |
| 1.3500 | -.0025 | -.111 | -.0627 | .3515  | .0452  | -.0183 | -.0185 |
| 2.0000 | -.0039 |       | -.0509 | -.2638 | -.0668 | -.0190 | -.0300 |
| 1.5000 | -.0028 | -.113 | -.0590 | .3398  | .0506  | -.0185 | -.0214 |
| 3.0000 | -.0044 |       | -.0220 | -.2483 | -.0739 | -.0184 | -.0343 |
| 2.2500 | -.0036 | -.115 | -.0319 | .3141  | .0617  | -.0177 | -.0278 |
| 4.0000 | -.0044 |       | -.0075 | -.2447 | -.0745 | -.0182 | -.0347 |
| 3.0000 | -.0037 | -.115 | -.0134 | .3084  | .0632  | -.0173 | -.0289 |
| 5.0000 | -.0044 |       | -.0023 | -.2434 | -.0745 | -.0182 | -.0347 |
| 3.7500 | -.0038 | -.115 | -.0050 | .3066  | .0634  | -.0172 | -.0290 |

TABLE B.26. NUMERICAL VALUES FOR FACTORS IN EQUS. (3.45) TO (3.50).

287

```
SPR......=  .100
SPL......=  .056
PL.......= 1.000
PR.......= 0.000
POIS.R...=  .330
K........=  .750
```

| RR / RL | KWR / KWL | KMC | KQXR / KQXL | KQYRN / KQYLN | KQYRF / KQYLF | KMXR / KMXL | KMYR / KMYL |
|---|---|---|---|---|---|---|---|
| 1.0000 | -.0004 |  | -.0166 | -.0842 | -.0072 | -.0006 | -.0063 |
| .7500 | -.0002 | -.046 | -.0134 | .1120 | .0034 | -.0010 | -.0031 |
| 1.2000 | -.0008 |  | -.0248 | -.1016 | -.0145 | -.0020 | -.0103 |
| .9000 | -.0004 | -.057 | -.0223 | .1344 | .0080 | -.0025 | -.0058 |
| 1.4000 | -.0012 |  | -.0307 | -.1110 | -.0225 | -.0036 | -.0143 |
| 1.0500 | -.0008 | -.066 | -.0297 | .1455 | .0141 | -.0042 | -.0088 |
| 1.6000 | -.0017 |  | -.0329 | -.1150 | .0299 | -.0050 | -.0178 |
| 1.2000 | -.0011 | -.071 | -.0348 | .1492 | .0204 | -.0057 | -.0117 |
| 1.8000 | -.0021 |  | -.0375 | -.1159 | -.0360 | -.0060 | -.0206 |
| 1.3500 | -.0014 | -.075 | -.0340 | .1487 | .0262 | -.0068 | -.0144 |
| 2.0000 | -.0024 |  | -.0381 | -.1153 | -.0407 | -.0068 | -.0227 |
| 1.5000 | -.0017 | -.077 | -.0210 | .1464 | .0311 | -.0076 | -.0166 |
| 3.0000 | -.0030 |  | -.0274 | -.1109 | .0501 | -.0079 | -.0269 |
| 2.2500 | -.0025 | -.080 | -.0094 | .1358 | .0430 | -.0087 | -.0222 |
| 4.0000 | -.0031 |  | -.0142 | -.1099 | -.0512 | -.0079 | -.0274 |
| 3.0000 | -.0027 | -.080 | -.0036 | .1333 | .0450 | -.0086 | -.0233 |
| 5.0000 | -.0031 |  | -.0062 | -.1097 | -.0513 | -.0079 | -.0275 |
| 3.7500 | -.0027 | -.080 |  | .1330 | .0453 | -.0085 | -.0235 |

TABLE B.27. NUMERICAL VALUES FOR FACTORS IN EQUS. (3.45) TO (3.50).

SPR......= .200
SPL......= .113
PL.......= 1.000
PR.......= 0.000
POIS.R...= .300
K........= .750

| RR / RL | KWR / KWL | KMC | KQXR / KQXL | KQYRN / KQYLN | KQYRF / KQYLF | KMXR / KMXL | KMYR / KMYL |
|---|---|---|---|---|---|---|---|
| 1.0000 | -.0002 | | -.0099 | -.0505 | -.0043 | .0018 | -.0059 |
| .7500 | -.0001 | -.039 | -.0061 | .0671 | .0020 | .0007 | -.0032 |
| 1.2000 | -.0005 | | -.0156 | -.0638 | -.0091 | .0016 | -.0094 |
| .9000 | -.0003 | -.048 | -.0140 | .0844 | .0051 | .0003 | -.0056 |
| 1.4000 | -.0008 | | -.0202 | -.0726 | -.0148 | .0008 | -.0126 |
| 1.0500 | -.0005 | -.055 | -.0196 | .0951 | .0093 | -.0004 | -.0081 |
| 1.6000 | -.0011 | | -.0232 | -.0776 | -.0204 | -.0001 | -.0154 |
| 1.2000 | -.0008 | -.060 | -.0238 | .1006 | .0139 | -.0014 | -.0105 |
| 1.8000 | -.0014 | | -.0247 | -.0801 | -.0254 | -.0011 | -.0176 |
| 1.3530 | -.0010 | -.063 | -.0265 | .1027 | .0185 | -.0023 | -.0126 |
| 2.0000 | -.0017 | | -.0248 | -.0811 | -.0294 | -.0019 | -.0193 |
| 1.5000 | -.0013 | -.065 | -.0277 | .1327 | .0225 | -.0031 | -.0144 |
| 3.0000 | -.0023 | | -.0173 | -.0799 | -.0387 | -.0038 | -.0229 |
| 2.2500 | -.0019 | -.067 | -.0221 | .0971 | .0333 | -.0049 | -.0189 |
| 4.0000 | -.0024 | | -.0088 | -.0790 | -.0401 | -.0041 | -.0234 |
| 3.0000 | -.0021 | -.068 | -.0127 | .0949 | .0356 | -.0051 | -.0199 |
| 5.0000 | -.0024 | | -.0038 | -.0789 | -.0403 | -.0041 | -.0235 |
| 3.7500 | -.0022 | -.068 | -.0062 | .0945 | .0360 | -.0051 | -.0201 |

TABLE B.28. NUMERICAL VALUES FOR FACTORS IN EQUS. (3.45) TO (3.50).

289

SPR......= .300
SPL......= .169
PL.......= 1.000
PR.......= 0.000
POIS.R...= .300
K........= .750

| RR / RL | KWR / KWL | KMC | KQXR / KQXL | KQYRN / KQYLN | KQYRF / KQYLF | KMXR / KMXL | KMYR / KMYL |
|---|---|---|---|---|---|---|---|
| 1.0000 | -.0002 | | -.0071 | -.0360 | -.0031 | .0029 | -.0058 |
| .7500 | -.0001 | -.036 | -.0058 | -.0479 | .0014 | .0015 | -.0032 |
| 1.2000 | -.0004 | | -.0114 | -.0465 | -.0067 | .0033 | -.0089 |
| .9000 | -.0002 | -.044 | -.0102 | .0615 | .0037 | .0017 | -.0055 |
| 1.4000 | -.0006 | | -.0151 | -.0540 | -.0110 | .0030 | -.0118 |
| 1.0500 | -.0004 | -.050 | -.0146 | -.0707 | .0069 | .0014 | -.0078 |
| 1.6000 | -.0009 | | -.0177 | -.0587 | -.0155 | .0024 | -.0142 |
| 1.2800 | -.0006 | -.053 | -.0181 | .0761 | .0106 | .0008 | -.0099 |
| 1.8000 | -.011 | | -.0191 | -.0614 | -.0196 | .0016 | -.0161 |
| 1.3500 | -.0008 | -.056 | -.0205 | .0787 | .0143 | .0001 | -.0117 |
| 2.0000 | -.0013 | | -.0195 | -.0628 | -.0230 | .0008 | -.0175 |
| 1.5000 | -.0010 | -.058 | -.0217 | .0795 | .0176 | -.0006 | -.0132 |
| 3.0000 | -.0019 | | -.0145 | -.0632 | -.0315 | -.0014 | -.0203 |
| 2.2500 | -.0016 | -.060 | -.0184 | .0766 | .0273 | -.0026 | -.0169 |
| 4.0000 | -.020 | | -.0079 | -.0627 | -.0331 | -.0019 | -.0208 |
| 3.0000 | -.0018 | -.060 | -.0111 | .0748 | .0296 | -.0030 | -.0177 |
| 5.0000 | -.0020 | | -.0037 | -.0625 | -.0334 | -.0020 | -.0208 |
| 3.7500 | -.0018 | -.060 | -.0057 | .0744 | .0300 | -.0031 | -.0178 |

TABLE B.29. NUMERICAL VALUES FOR FACTORS IN EQUS. (3.45) TO (3.50).

```
SPR......= .400
SPL......= .225
PL.......= 1.000
PR.......= 0.000
POIS.R...= .300
K........= .750
```

| RR / RL | KWR / KWL | KMC | KQXR / KQXL | KQYRN / KQYLN | KQYRF / KQYLF | KMXR / KMXL | KMYR / KMYL |
|---|---|---|---|---|---|---|---|
| 1.0000 | -.0001 |  | -.0055 | -.0280 | -.0024 | .0034 | -.0057 |
| .7500 | -.0001 | -.034 | -.0045 | .0372 | .0011 | .0019 | -.0033 |
| 1.2000 | -.0003 |  | -.0090 | -.0366 | -.0052 | .0042 | -.0087 |
| .9000 | -.0002 | -.041 | -.0081 | -.0484 | -.0029 | .0024 | -.0054 |
| 1.4000 | -.0005 |  | -.0120 | -.0430 | .0088 | .0043 | -.0113 |
| 1.0500 | -.0003 | -.046 | -.0116 | .0563 | .0055 | .0025 | -.0076 |
| 1.6000 | -.0007 |  | -.0142 | -.0472 | -.0125 | .0039 | -.0134 |
| 1.2000 | -.0005 | -.050 | -.0146 | .0612 | .0085 | .0022 | -.0095 |
| 1.8000 | -.0009 |  | -.0156 | -.0498 | -.0160 | .0033 | -.0151 |
| 1.3500 | -.0006 | -.052 | -.0167 | .0638 | .0116 | .0017 | -.0111 |
| 2.0000 | -.0011 |  | -.0161 | -.0513 | -.0189 | .0026 | -.0162 |
| 1.5000 | -.0008 | -.053 | -.0179 | .0650 | .0145 | .0011 | -.0123 |
| 3.0000 | -.0016 |  | -.0124 | -.0525 | -.0267 | .0002 | -.0186 |
| 2.2500 | -.0013 | -.055 | -.0157 | .0635 | .0231 | -.0010 | -.0154 |
| 4.0000 | -.0017 |  | -.0070 | -.0522 | -.0283 | -.0004 | -.0189 |
| 3.0000 | -.0015 | -.055 | -.0098 | .0621 | .0253 | -.0016 | -.0161 |
| 5.0000 | -.0017 |  | -.0035 | -.0521 | -.0285 | -.0005 | -.0189 |
| 3.7500 | -.0015 | -.055 | -.0053 | .0617 | .0257 | .0017 | .0162 |

TABLE B.30. NUMERICAL VALUES FOR FACTORS IN EQUS. (3.45) TO (3.50).

SPR...... = .500
SPL...... = .281
PL....... = 1.000
PR....... = 0.000
POIS.R... = .300
K........ = .750

| RR RL | KWR KWL | KMC | KQXR KQXL | KQYRN KQYLN | KQYRF KQYLF | KMXR KMXL | KMYR KMYL |
|---|---|---|---|---|---|---|---|
| 1.0000 | -.0001 | | -.0045 | -.0229 | -.0020 | .0038 | -.0057 |
| .7500 | -.0001 | -.033 | -.0037 | -.0305 | -.0069 | .0021 | -.0033 |
| 1.2000 | -.0002 | | -.0074 | -.0302 | -.0043 | .0048 | -.0085 |
| .9000 | -.0001 | -.046 | -.0067 | -.0399 | -.0024 | .0029 | -.0054 |
| 1.4000 | -.0004 | | -.0100 | -.0357 | -.0373 | .0052 | -.0110 |
| 1.0500 | -.0002 | -.044 | -.0097 | -.0468 | -.0046 | .0032 | -.0074 |
| 1.6000 | -.0006 | | -.0149 | -.0395 | -.0105 | .0050 | -.0129 |
| 1.2000 | -.0004 | -.047 | -.0122 | -.0512 | -.0071 | .0031 | -.0092 |
| 1.8000 | -.0008 | | -.0131 | -.0419 | -.0135 | .0044 | -.0144 |
| 1.3500 | -.0005 | -.049 | -.0141 | -.0537 | -.0098 | .0027 | -.0106 |
| 2.0000 | -.0009 | | -.0137 | -.0434 | -.0161 | .0038 | -.0154 |
| 1.5000 | -.0007 | -.050 | -.0152 | -.0549 | -.0123 | .0022 | -.0118 |
| 3.0000 | -.0014 | | -.0199 | -.0450 | -.0231 | .0014 | -.0173 |
| 2.2500 | -.0012 | -.051 | -.0137 | -.0544 | -.0206 | .0001 | -.0144 |
| 4.0000 | -.0015 | | -.0063 | -.0448 | -.0247 | .0007 | -.0175 |
| 3.0000 | -.0013 | -.051 | -.0088 | -.0532 | -.0221 | -.0006 | -.0149 |
| 5.0000 | -.0015 | | -.0032 | -.0447 | -.0250 | -.0005 | -.0175 |
| 3.7500 | -.0014 | -.051 | -.0048 | -.0529 | -.0226 | -.0007 | -.0150 |

TABLE B.31. NUMERICAL VALUES FOR FACTORS IN EQUS. (3.45) TO (3.50).

```
SPR......= 0.000
SPL......= 0.000
PL.......= 0.000
PR.......= 1.000
POIS.R...= .300
K........= .750
```

| RR / RL | KWR / KWL | KMC | KQXR / KQXL | KQYRN / KQYLN | KQYRF / KQYLF | KMXR / KMXL | KMYR / KMYL |
|---|---|---|---|---|---|---|---|
| 1.0000 | -.0013 |       | -.0575 | -.3824 | -.0252 | -.0149 | -.0091 |
| .7500  | -.0006 | -.098 | -.0469 | .5086  | .0117  | -.0114 | -.0032 |
| 1.2000 | -.0019 |       | -.0602 | -.3521 | -.0359 | -.0166 | -.0141 |
| .9000  | -.0011 | -.103 | -.0547 | .4666  | .0199  | -.0139 | -.0067 |
| 1.4000 | -.0024 |       | -.0565 | -.3292 | -.0434 | -.0169 | -.0180 |
| 1.0500 | -.0015 | -.104 | -.0560 | .4341  | .0270  | -.0151 | -.0100 |
| 1.6000 | -.0027 |       | -.0456 | -.3138 | -.0480 | -.0167 | -.0205 |
| 1.2000 | -.0018 | -.104 | -.0532 | .4119  | .0325  | -.0153 | -.0127 |
| 1.8000 | -.0029 |       | -.0416 | -.3040 | -.0506 | -.0162 | -.0221 |
| 1.3500 | -.0020 | -.103 | -.0480 | .3973  | .0363  | -.0152 | -.0147 |
| 2.0000 | -.0030 |       | -.0338 | -.2979 | -.0519 | -.0158 | -.0230 |
| 1.5000 | -.0022 | -.103 | -.0418 | .3881  | .0388  | -.0148 | -.0162 |
| 3.0000 | -.0031 |       | -.0084 | -.2885 | -.0526 | -.0150 | -.0239 |
| 2.2500 | -.0025 | -.101 | -.0155 | .3733  | .0423  | -.0137 | -.0186 |
| 4.0000 | -.0031 |       | -.0011 | -.2866 | -.0525 | -.0150 | -.0238 |
| 3.0000 | -.0025 | -.101 | -.0042 | .3705  | .0425  | -.0135 | -.0189 |
| 5.0000 | -.0031 |       | .0061  | -.2852 | -.0525 | -.0150 | -.0238 |
| 3.7500 | -.0025 | -.101 | -.0009 | .3686  | .0425  | -.0134 | -.0189 |

TABLE B.32. NUMERICAL VALUES FOR FACTORS IN EQUS. (3.45) TO (3.50).

SPR...... = -.100
SPL...... = .056
PL....... = 0.000
PR....... = 1.000
POIS.R... = .300
K........ = .750

| RR RL | KWR KWL | KMC | KQXR KQXL | KQYRN KQYLN | KQYRF KQYLF | KHXR KHXL | KMYR KMYL |
|---|---|---|---|---|---|---|---|
| 1.0000 | -.0004 | | -.0193 | -.0996 | -.0084 | -.0007 | -.0073 |
| .7500 | -.0002 | -.054 | -.0156 | .1324 | -.0039 | -.0012 | -.0037 |
| 1.2000 | -.0008 | | -.0253 | -.1072 | -.0148 | -.0021 | -.0106 |
| .9000 | -.0004 | -.061 | -.0228 | .1417 | .0082 | -.0026 | -.0059 |
| 1.4000 | -.0011 | | -.0262 | -.1080 | -.0208 | -.0033 | -.0133 |
| 1.0500 | -.0007 | -.064 | -.0274 | .1418 | .0130 | -.0039 | -.0081 |
| 1.6000 | -.0014 | | -.0286 | -.1060 | -.0256 | -.0042 | -.0153 |
| 1.2000 | -.0009 | -.066 | -.0296 | .1379 | .0174 | -.0048 | -.0101 |
| 1.8000 | -.0016 | | -.0273 | -.1033 | -.0289 | -.0047 | -.0167 |
| 1.3500 | -.0012 | -.066 | -.0297 | .1331 | .0210 | -.0055 | -.0116 |
| 2.0000 | -.0018 | | -.0250 | -.1009 | -.0312 | -.0050 | -.0176 |
| 1.5000 | -.0013 | -.066 | -.0284 | .1289 | .0238 | -.0058 | -.0128 |
| 3.0000 | -.0020 | | -.0114 | -.0959 | -.0342 | -.0052 | -.0189 |
| 2.2500 | -.0017 | -.066 | -.0159 | .1199 | .0288 | -.0059 | -.0152 |
| 4.0000 | -.0020 | | -.0035 | -.0955 | -.0342 | -.0051 | -.0189 |
| 3.0000 | -.0017 | -.065 | -.0064 | .1191 | .0293 | -.0057 | -.0155 |
| 5.0000 | -.0020 | | -.0007 | -.0955 | -.0342 | -.0051 | -.0189 |
| 3.7500 | -.0017 | -.065 | -.0021 | .1190 | .0293 | -.0057 | -.0155 |

TABLE B.33. NUMERICAL VALUES FOR FACTORS IN EQUS. (3.45) TO (3.50).

SPR......= .200
SPL......= .113
PL.......= 0.000
PR.......= 1.000
POIS.R...= .300
K........= .750

| RR / RL | KWR / KWL | KMC | KQXR / KQXL | KQYRN / KQYLN | KQYRF / KQYLF | KMXR / KMXL | KHYR / KMYL |
|---|---|---|---|---|---|---|---|
| 1.0000 | -.0003 | | -.0116 | -.0595 | -.0050 | .0021 | -.0069 |
| .7500 | -.0001 | -.046 | -.0094 | .0791 | .0023 | .0008 | -.0037 |
| 1.2000 | -.0005 | | -.0160 | -.0670 | -.0093 | .0017 | -.0096 |
| .9000 | -.0003 | -.051 | -.0144 | -.0886 | -.0052 | -.0004 | -.0057 |
| 1.4000 | -.0008 | | -.0186 | -.0699 | -.0137 | -.0008 | -.0118 |
| 1.0500 | -.0005 | -.054 | -.0181 | .0918 | .0086 | -.0004 | -.0075 |
| 1.6000 | -.0010 | | -.0197 | -.0704 | -.0175 | -.0000 | -.0133 |
| 1.2000 | -.0006 | -.055 | -.0202 | .0915 | .0119 | -.0011 | -.0090 |
| 1.8000 | -.0012 | | -.0194 | -.0698 | -.0204 | -.0007 | -.0144 |
| 1.3500 | -.0008 | -.056 | -.0210 | -.0898 | .0148 | -.0018 | -.0102 |
| 2.0000 | -.0013 | | -.0184 | -.0688 | -.0225 | -.0012 | -.0151 |
| 1.5000 | -.0010 | -.056 | -.0268 | .0877 | .0172 | -.0023 | -.0111 |
| 3.0000 | -.016 | | -.0099 | -.0657 | -.0262 | -.0021 | -.0162 |
| 2.2500 | -.0013 | -.056 | -.0133 | .0814 | .0223 | -.0031 | -.0131 |
| 4.0000 | -.0016 | | -.0039 | -.0652 | -.0264 | -.0021 | -.0163 |
| 3.0000 | -.0014 | -.055 | -.0062 | .0804 | .0229 | -.0030 | -.0134 |
| 5.0000 | -.0016 | | -.0012 | -.0652 | -.0264 | -.0021 | -.0163 |
| 3.7500 | -.0014 | -.055 | -.0024 | .0803 | .0230 | -.0030 | -.0134 |

TABLE B.34. NUMERICAL VALUES FOR FACTORS IN EQUS. (3.45) TO (3.50).

```
SPR......=  .300
SPL......=  .169
PL.......= 0.000
PR.......= 1.000
POIS.R...=  .300
K........=  .750
```

| RR / RL | KHR / KHL | KMC | KQXR / KQXL | KQYRN / KQYLN | KQYRF / KQYLF | KMXR / KMXL | KMYR / KMYL |
|---|---|---|---|---|---|---|---|
| 1.0000 | -.0002 |        | -.0083 | -.0425 | -.0036 |  .0033 | -.0068 |
|  .7500 | -.0001 | -.043  | -.0067 | -.0565 | -.0017 |  .0017 | -.0038 |
| 1.2000 | -.0004 |        | -.0117 | -.0487 | -.0068 |  .0034 | -.0092 |
|  .9000 | -.0002 | -.047  | -.0105 |  .0644 | -.0038 |  .0017 | -.0056 |
| 1.4000 | -.0006 |        | -.0139 | -.0518 | -.0102 |  .0029 | -.0110 |
| 1.0500 | -.0003 | -.049  | -.0135 |  .0680 |  .0064 |  .0013 | -.0072 |
| 1.6000 | -.0007 |        | -.0150 | -.0529 | -.0133 |  .0022 | -.0123 |
| 1.2000 | -.0005 | -.050  | -.0154 | -.0688 | -.0090 |  .0008 | -.0085 |
| 1.8000 | -.0009 |        | -.0151 | -.0530 | -.0157 |  .0015 | -.0131 |
| 1.3500 | -.0006 | -.050  | -.0163 |  .0682 |  .0114 |  .0002 | -.0095 |
| 2.0000 | -.0010 |        | -.0145 | -.0527 | -.0176 |  .0009 | -.0137 |
| 1.5000 | -.0007 | -.050  | -.0163 |  .0670 |  .0134 | -.0003 | -.0102 |
| 3.0000 | -.0013 |        | -.0085 | -.0508 | -.0213 | -.0003 | -.0146 |
| 2.2500 | -.0011 | -.050  | -.0112 | -.0627 |  .0182 | -.0014 | -.0117 |
| 4.0000 | -.0013 |        | -.0037 | -.0504 | -.0217 | -.0004 | -.0146 |
| 3.0000 | -.0011 | -.050  | -.0057 |  .0617 |  .0190 | -.0015 | -.0120 |
| 5.0000 | -.0013 |        | -.0014 | -.0504 | -.0217 | -.0004 | -.0146 |
| 3.7500 | -.0011 | -.050  | -.0024 |  .0616 |  .0190 | -.0015 | -.0120 |

TABLE B.35. NUMERICAL VALUES FOR FACTORS IN EQUS. (3.45) TO (3.50).

```
SPR....... =  .400
SPL....... =  .225
PL........ = 0.000
PR........ = 1.000
POIS.R... =  .300
K......... =  .750
```

| RR RL | KWR KWL | KMC | KQXR KQXL | KQYRN KQYLN | KQYRF KQYLF | KMXR KMXL | KMYR KMYL |
|---|---|---|---|---|---|---|---|
| 1.0000 | -.0001 |       | -.0064 | -.0330 | -.0028 | .0040 | -.0067 |
|  .7500 | -.0001 | -.041 | -.0052 |  .0439 |  .0013 | .0022 | -.0038 |
| 1.2000 | -.0003 |       | -.0052 | -.0383 | -.0054 | .0044 | -.0089 |
|  .9000 | -.0002 | -.044 | -.0083 |  .0507 | -.0030 | .0025 | -.0056 |
| 1.4000 | -.0004 |       | -.0131 | -.0412 | -.0081 | .0041 | -.0106 |
| 1.0500 | -.0003 | -.046 | -.0107 |  .0540 |  .0051 | .0023 | -.0070 |
| 1.6000 | -.0006 |       | -.0121 | -.0424 | -.0137 | .0035 | -.0116 |
| 1.2000 | -.0004 | -.047 | -.0124 |  .0551 |  .0073 | .0019 | -.0081 |
| 1.6000 | -.0007 |       | -.0123 | -.0428 | -.0128 | .0029 | -.0123 |
| 1.3500 | -.0005 | -.047 | -.0133 |  .0550 |  .0093 | .0014 | -.0090 |
| 2.0800 | -.0008 |       | -.0120 | -.0428 | -.0145 | .0023 | -.0128 |
| 1.5000 | -.0006 | -.047 | -.0134 |  .0544 |  .0111 | .0010 | -.0096 |
| 3.0000 | -.0011 |       | -.0073 | -.0417 | -.0179 | .0009 | -.0134 |
| 2.2500 | -.0009 | -.046 | -.0096 |  .0513 |  .0154 | .0003 | -.0108 |
| 4.0000 | -.0011 |       | -.0034 | -.0414 | -.0184 | .0007 | -.0134 |
| 3.0000 | -.0010 | -.046 | -.0051 |  .0504 |  .0162 | .0005 | -.0110 |
| 5.0000 | -.0011 |       | -.0014 | -.0413 | -.0185 | .0007 | -.0134 |
| 3.7500 | -.0010 | -.046 | -.0023 |  .0503 |  .0163 | .0005 | -.0110 |

TABLE B.36. NUMERICAL VALUES FOR FACTORS IN EQUS. (3.45) TO (3.50).

SPR......= .500
SPL......= .281
PL......= 0.000
PR......= 1.000
POIS.R...= .300
K.......= .750

| RR RL | KWR KWL | KMC | KQXR KQXL | KQYRN KQYLN | KQYRF KQYLF | KMXR KMXL | KMYR KMYL |
|---|---|---|---|---|---|---|---|
| 1.0000 | -.0001 | | -.0053 | -.0270 | -.0023 | .0044 | -.0066 |
| .7500 | -.0001 | -.039 | -.0043 | .0359 | .0011 | .0025 | -.0038 |
| 1.2000 | -.0002 | | -.0076 | -.0316 | -.0044 | .0050 | -.0088 |
| .9000 | -.0001 | -.043 | -.0068 | .0417 | .0025 | .0030 | -.0055 |
| 1.4000 | -.004 | | -.0002 | -.0342 | -.0068 | .0049 | -.0103 |
| 1.0500 | -.002 | -.044 | -.0089 | .0448 | .0042 | .0030 | -.0069 |
| 1.6000 | -.005 | | -.0101 | -.0354 | -.0090 | .0044 | -.0112 |
| 1.2000 | -.003 | -.044 | -.0104 | .0460 | .0061 | .0027 | -.0079 |
| 1.8000 | -.0006 | | -.0108 | -.0359 | -.0108 | .0038 | -.0118 |
| 1.3500 | -.0004 | -.045 | -.0112 | .0462 | .0079 | .0023 | -.0086 |
| 2.0000 | -.0007 | | -.0162 | -.0360 | -.0123 | .0032 | -.0121 |
| 1.5000 | -.0005 | -.044 | -.0114 | .0458 | .0094 | .0018 | -.0091 |
| 3.0000 | -.0009 | | -.0065 | -.0354 | -.0155 | .0017 | -.0125 |
| 2.2500 | -.0008 | -.044 | -.00F4 | -.0435 | .0133 | .0005 | -.0101 |
| 4.0000 | -.0010 | | -.0031 | -.0352 | -.0160 | .0015 | -.0125 |
| 3.0000 | -.0008 | -.043 | -.0046 | .0428 | .0141 | .0002 | -.0102 |
| 5.0000 | -.0010 | | -.0013 | -.0351 | -.0161 | .0014 | -.0125 |
| 3.7500 | -.0009 | -.043 | -.0022 | .0426 | .0142 | .0002 | -.0102 |

TABLE B.37. NUMERICAL VALUES FOR FACTORS IN EQUS. (3.45) TO (3.50).

SPR...... = 0.000
SPL...... = 0.000
PL....... = 1.000
PR....... = 0.000
POIS.R... = .300
K........ = 1.000

| RR RL | KWR KWL | KMC | KQXR KQXL | KQYRN KQYLN | KQYRF KQYLF | KMXR KMXL | KMYR KMYL |
|---|---|---|---|---|---|---|---|
| 1.0000 | -.0014 |  | -.0580 | -.3848 | -.0254 | -.0150 | -.0092 |
| 1.0000 | -.0014 | -.099 | -.0580 | -.3848 | -.0254 | -.0150 | -.0092 |
| 1.2000 | -.0020 |  | -.0613 | -.3566 | -.0365 | -.0169 | -.0144 |
| 1.2000 | -.0020 | -.105 | -.0613 | -.3566 | -.0365 | -.0169 | -.0144 |
| 1.4000 | -.0025 |  | -.0583 | -.3354 | -.0447 | -.0174 | -.0185 |
| 1.4000 | -.0025 | -.107 | -.0583 | -.3354 | -.0447 | -.0174 | -.0185 |
| 1.6000 | -.0028 |  | -.0519 | -.3212 | -.0500 | -.0173 | -.0214 |
| 1.6000 | -.0028 | -.108 | -.0519 | -.3212 | -.0500 | -.0173 | -.0214 |
| 1.8000 | -.0030 |  | -.0443 | -.3121 | -.0533 | -.0170 | -.0233 |
| 1.8000 | -.0030 | -.108 | -.0443 | -.3121 | -.0533 | -.0170 | -.0233 |
| 2.0000 | -.0032 |  | -.0367 | -.3064 | -.0552 | -.0167 | -.0245 |
| 2.0000 | -.0032 | -.108 | -.0367 | -.3064 | -.0552 | -.0167 | -.0245 |
| 3.0000 | -.0034 |  | -.0108 | -.2969 | -.0574 | -.0160 | -.0261 |
| 3.0000 | -.0034 | -.107 | -.0108 | -.2969 | -.0574 | -.0160 | -.0261 |
| 4.0000 | -.0034 |  | -.0025 | -.2948 | -.0575 | -.0160 | -.0262 |
| 4.0000 | -.0034 | -.107 | -.0025 | -.2948 | -.0575 | -.0160 | -.0262 |
| 5.0000 | -.0034 |  | -.0005 | -.2934 | -.0575 | -.0160 | -.0262 |
| 5.0000 | -.0034 | -.107 | -.0005 | -.2934 | -.0575 | -.0160 | -.0262 |

TABLE B.38. NUMERICAL VALUES FOR FACTORS IN EQUS. (3.45) TO (3.50).

299

```
SPR.......= .100
SPL.......= .100
PL........= 1.000
PR........= 0.000
POIS.R....= .300
K.........= 1.000
```

| RR / RL | KWR / KWL | KMC | KQXR / KQXL | KQYRN / KQYLN | KQYRF / KQYLF | KMXR / KMXL | KMYR / KMYL |
|---|---|---|---|---|---|---|---|
| 1.0000 | -.0004 |        | -.0193 | -.0998 | -.0084 | -.0007 | -.0073 |
| 1.0000 | -.0004 | -.054  | -.0193 |  .0998 |  .0084 | -.0007 | -.0073 |
| 1.2000 | -.0008 |        | -.0254 | -.1077 | -.0149 | -.0021 | -.0106 |
| 1.2000 | -.0008 | -.061  | -.0254 |  .1077 |  .0149 | -.0021 | -.0106 |
| 1.4000 | -.0012 |        | -.0285 | -.1090 | -.0210 | -.0033 | -.0134 |
| 1.4000 | -.0012 | -.065  | -.0285 |  .1090 |  .0210 | -.0033 | -.0134 |
| 1.6000 | -.0015 |        | -.0291 | -.1075 | -.0260 | -.0043 | -.3156 |
| 1.6000 | -.0015 | -.067  | -.0291 |  .1075 |  .0260 | -.0043 | -.0156 |
| 1.8000 | -.0017 |        | -.0279 | -.1052 | -.0296 | -.0048 | -.0171 |
| 1.8000 | -.0017 | -.068  | -.0279 |  .1052 |  .0296 | -.0048 | -.0181 |
| 2.0000 | -.0019 |        | -.0257 | -.1031 | -.0321 | -.0052 | -.0181 |
| 2.0000 | -.0019 | -.068  | -.0257 |  .1031 |  .0321 | -.0052 | -.0198 |
| 3.0000 | -.0021 |        | -.0123 | -.0988 | -.0359 | -.0055 | -.0198 |
| 3.0000 | -.0021 | -.068  | -.0123 |  .0988 |  .0359 | -.0055 | -.0200 |
| 4.0000 | -.0022 |        | -.0043 | -.0984 | -.0362 | -.0055 | -.0200 |
| 4.0000 | -.0022 | -.068  | -.0043 |  .0984 |  .0362 | -.0055 | -.0200 |
| 5.0000 | -.0022 |        | -.0012 | -.0984 | -.0362 | -.0054 | -.0200 |
| 5.0000 | -.0022 | -.068  | -.0012 |  .0984 |  .0362 | -.0054 | -.0200 |

TABLE B.39. NUMERICAL VALUES FOR FACTORS IN EQUS. (3.45) TO (3.50).

```
SPR....... =  .200
SPL....... =  .200
PL........ = 1.000
PR........ = 0.000
POIS.R... =  .300
K......... = 1.000
```

| RR RL | KHR KWL | KMC | KQXR KQXL | KQYRN KQYLN | KQYRF KQYLF | KMXR KMXL | KMYR KMYL |
|---|---|---|---|---|---|---|---|
| 1.0000 | -.0003 |       | -.0116 | -.0596 | -.0050 |  .0021 | -.0069 |
| 1.0000 | -.0003 | -.046 | -.0116 | -.0596 | -.0050 |  .0021 | -.0069 |
| 1.2000 | -.0005 |       | -.0160 | -.0671 | -.0093 |  .0017 | -.0096 |
| 1.2000 | -.0005 | -.051 | -.0160 | -.0671 | -.0093 |  .0017 | -.0096 |
| 1.4000 | -.0008 |       | -.0187 | -.0702 | -.0137 |  .0008 | -.0118 |
| 1.4000 | -.0008 | -.054 | -.0187 | -.0702 | -.0137 |  .0008 | -.0118 |
| 1.6000 | -.0010 |       | -.0198 | -.0708 | -.0176 |  .0000 | -.0134 |
| 1.6000 | -.0010 | -.056 | -.0198 | -.0708 | -.0176 |  .0000 | -.0134 |
| 1.8000 | -.0012 |       | -.0196 | -.0703 | -.0206 | -.0007 | -.0145 |
| 1.8000 | -.0012 | -.056 | -.0196 | -.0703 | -.0206 | -.0007 | -.0145 |
| 2.0000 | -.0013 |       | -.0186 | -.0695 | -.0228 | -.0012 | -.0153 |
| 2.0000 | -.0013 | -.057 | -.0186 | -.0695 | -.0228 | -.0012 | -.0153 |
| 3.0000 | -.0016 |       | -.0103 | -.0668 | -.0269 | -.0021 | -.0166 |
| 3.0000 | -.0016 | -.057 | -.0103 | -.0668 | -.0269 | -.0021 | -.0166 |
| 4.0000 | -.0016 |       | -.0042 | -.0664 | -.0273 | -.0022 | -.0168 |
| 4.0000 | -.0016 | -.057 | -.0042 | -.0664 | -.0273 | -.0022 | -.0168 |
| 5.0000 | -.0016 |       | -.0015 | -.0664 | -.0273 | -.0022 | -.0168 |
| 5.0000 | -.0016 | -.056 | -.0015 | -.0664 | -.0273 | -.0022 | -.0168 |

TABLE B.40. NUMERICAL VALUES FOR FACTORS IN EQUS. (3.45) TO (3.50).

301

```
SPR......= .300
SPL......= .300
PL......= 1.000
PR......= 0.000
POIS.R...= .300
K......= 1.000
```

| RR / RL | KWR / KWL | KMC | KQXR / KQXL | KQYRN / KQYLN | KQYRF / KQYLF | KMXR / KMXL | KMYR / KMYL |
|---|---|---|---|---|---|---|---|
| 1.0800 | -.0002 |  | -.0083 | -.0425 | -.0036 | .0033 | -.0068 |
| 1.0000 | -.0002 | -.043 | -.0083 | .0425 | .0036 | .0033 | -.0068 |
| 1.2000 | -.0004 |  | -.0117 | -.0487 | -.0068 | .0034 | -.0092 |
| 1.2000 | -.0004 | -.047 | -.0117 | .0487 | .0068 | .0034 | -.0092 |
| 1.4000 | -.0006 |  | -.0139 | -.0518 | -.0102 | .0029 | -.0110 |
| 1.4000 | -.0006 | -.049 | -.0139 | .0518 | .0102 | .0029 | -.0110 |
| 1.6000 | -.0007 |  | -.0150 | -.0530 | -.0133 | .0022 | -.0123 |
| 1.6000 | -.0007 | -.050 | -.0150 | .0530 | .0133 | .0022 | -.0123 |
| 1.8000 | -.0009 |  | -.0151 | -.0531 | -.0158 | .0015 | -.0132 |
| 1.8000 | -.0009 | -.050 | -.0151 | .0531 | .0158 | .0015 | -.0132 |
| 2.0000 | -.0010 |  | -.0145 | -.0528 | -.0177 | .0009 | -.0137 |
| 2.0000 | -.0010 | -.051 | -.0145 | .0528 | .0177 | .0009 | -.0137 |
| 3.0000 | -.0013 |  | -.0086 | -.0512 | -.0215 | -.0003 | -.0147 |
| 3.0000 | -.0013 | -.050 | -.0046 | .0512 | .0215 | -.0003 | -.0147 |
| 4.0000 | -.0013 |  | -.0038 | -.0508 | -.0220 | -.0005 | -.0148 |
| 4.0000 | -.0013 | -.050 | -.0038 | .0508 | .0220 | -.0005 | -.0148 |
| 5.0000 | -.0013 |  | -.0015 | -.0508 | -.0220 | -.0005 | -.0148 |
| 5.0000 | -.0013 | -.050 | -.0015 | .0508 | .0220 | -.0005 | -.0148 |

TABLE B.41. NUMERICAL VALUES FOR FACTORS IN EQUS. (3.45) TO (3.50).

302

SPR.......= .400
SPL.......= .400
PL........= 1.000
PR........= 0.000
POIS.R...= .300
K.........= 1.000

| RR RL | KWR KWL | KMC | KQXR KQXL | KQYRN KQYLN | KQYRF KQYLF | KMXR KMXL | KMYR KMYL |
|---|---|---|---|---|---|---|---|
| 1.0000 | -.0001 |  | -.0064 | -.0330 | -.0028 | .0040 | -.0067 |
| 1.0000 | -.0001 | -.041 | -.0064 | .0330 | .0028 | .0040 | -.0067 |
| 1.2000 | -.0003 |  | -.0052 | -.0383 | -.0054 | .0043 | -.0089 |
| 1.2000 | -.0003 | -.044 | -.0092 | .0383 | .0054 | .0043 | -.0089 |
| 1.4000 | -.0004 |  | -.0111 | -.0411 | -.0081 | .0041 | -.0105 |
| 1.6000 | -.0004 | -.046 | -.0111 | .0411 | .0081 | .0041 | -.0105 |
| 1.6000 | -.0006 |  | -.0120 | -.0424 | -.0107 | .0035 | -.0116 |
| 1.6000 | -.0006 | -.047 | -.0120 | .0424 | .0107 | .0035 | -.0116 |
| 1.8000 | -.0007 |  | -.0123 | -.0427 | -.0128 | .0029 | -.0123 |
| 1.8000 | -.0007 | -.047 | -.0123 | .0427 | .0128 | .0029 | -.0123 |
| 2.0000 | -.0008 |  | -.0119 | -.0427 | -.0144 | .0023 | -.0128 |
| 2.0000 | -.0008 | -.047 | -.0119 | .0427 | .0144 | .0023 | -.0128 |
| 3.0000 | -.0011 |  | -.0073 | -.0417 | -.0179 | .0009 | -.0134 |
| 3.0000 | -.0011 | -.046 | -.0073 | .0417 | .0179 | .0009 | -.0134 |
| 4.0000 | -.0011 |  | -.0034 | -.0414 | -.0184 | .0007 | -.0134 |
| 4.0000 | -.0011 | -.046 | -.0034 | .0414 | .0184 | .0007 | -.0134 |
| 5.0000 | -.0011 |  | -.0014 | -.0413 | -.0185 | .0007 | -.0134 |
| 5.0000 | -.0011 | -.046 | -.0014 | .0413 | .0185 | .0007 | -.0134 |

TABLE B.42. NUMERICAL VALUES FOR FACTORS IN EQUS. (3.45) TO (3.50).

SPR..... = .500
SPL..... = .500
PL...... = 1.000
PR...... = 0.000
POIS.R.. = .300
K....... = 1.000

| RR / RL | KWR / KWL | KMC | KQXR / KQXL | KQYRN / KQYLN | KQYRF / KQYLF | KMXR / KMXL | KMYR / KMYL |
|---|---|---|---|---|---|---|---|
| 1.0000 | -.0001 |      | -.0053 | -.0270 | -.0023 | .0044 | -.0066 |
| 1.0000 | -.0001 | -.039 | -.0053 | -.0270 | -.0023 | .0044 | -.0066 |
| 1.2000 | -.0002 |      | -.0076 | -.0315 | -.0044 | .0050 | -.0087 |
| 1.2000 | -.0002 | -.043 | -.0076 | -.0315 | -.0044 | .0050 | -.0087 |
| 1.4000 | -.0004 |      | -.0092 | -.0341 | -.0067 | .0049 | -.0102 |
| 1.4000 | -.0004 | -.044 | -.0092 | -.0341 | -.0067 | .0049 | -.0102 |
| 1.6000 | -.0005 |      | -.0111 | -.0353 | -.0089 | .0044 | -.0112 |
| 1.6000 | -.0005 | -.044 | -.0101 | -.0353 | -.0089 | .0044 | -.0112 |
| 1.8000 | -.0006 |      | -.0103 | -.0358 | -.0108 | .0038 | -.0117 |
| 1.8000 | -.0006 | -.044 | -.0103 | -.0358 | -.0108 | .0038 | -.0117 |
| 2.0000 | -.0007 |      | -.0101 | -.0359 | -.0122 | .0032 | -.0121 |
| 2.0000 | -.0007 | -.044 | -.0101 | -.0359 | -.0122 | .0032 | -.0121 |
| 3.0000 | -.0009 |      | -.0064 | -.0352 | -.0154 | .0017 | -.0124 |
| 3.0000 | -.0009 | -.043 | -.0064 | -.0352 | -.0154 | .0017 | -.0124 |
| 4.0000 | -.0010 |      | -.0031 | -.0350 | -.0159 | .0015 | -.0125 |
| 4.0000 | -.0010 | -.043 | -.0031 | -.0350 | -.0159 | .0015 | -.0125 |
| 5.0000 | -.0010 |      | -.0013 | -.0349 | -.0160 | .0014 | -.0125 |
| 5.0000 | -.0010 | -.043 | -.0013 | -.0349 | -.0160 | .0014 | -.0125 |

TABLE B.43. NUMERICAL VALUES FOR FACTORS IN EQUS. (3.45) TO (3.50).

```
SPR........ =  0.000
SPL........ =  0.000
FOL........ =  0.000
POR........ =  1.000
POIS.R..... =   .300
K.......... =   .250
```

| RR RL | KWR KWL | KMC | KQXR KQXL | KQYRN KQYLN | KQYRF KQYLF | KMXR KMXL | KMYR KMYL |
|---|---|---|---|---|---|---|---|
| 1.0000 | -.0006 |       | -.0274 | -.1200 | -.0118 | -.0069 | -.0043 |
|  .2500 | -.0000 | -.041 | -.0001 |  .0299 |  .0000 | -.0000 |  .0000 |
| 1.2000 | -.0010 |       | -.0339 | -.1133 | -.0194 | -.0088 | -.0077 |
|  .3000 | -.0000 | -.048 | -.0002 |  .0280 |  .0000 | -.0000 |  .0000 |
| 1.4000 | -.0015 |       | -.0376 | -.1032 | -.0265 | -.0098 | -.0111 |
|  .3500 | -.0000 | -.051 | -.0003 |  .0251 |  .0000 | -.0001 |  .0000 |
| 1.6000 | -.0018 |       | -.0393 | -.0922 | -.0324 | -.0103 | -.0141 |
|  .4000 | -.0000 | -.053 | -.0006 |  .0219 |  .0000 | -.0001 |  .0000 |
| 1.8000 | -.0021 |       | -.0398 | -.0817 | -.0369 | -.0103 | -.0165 |
|  .4500 | -.0000 | -.053 | -.0008 |  .0188 |  .0000 | -.0002 |  .0000 |
| 2.0000 | -.0023 |       | -.0397 | -.0724 | -.0402 | -.0100 | -.0184 |
|  .5000 | -.0000 | -.053 | -.0010 |  .0160 |  .0001 | -.0002 |  .0000 |
| 3.0000 | -.0026 |       | -.0369 | -.0439 | -.0434 | -.0076 | -.0211 |
|  .7500 | -.0000 | -.045 | -.0018 |  .0069 |  .0004 | -.0004 | -.0001 |
| 4.0000 | -.0024 |       | -.0354 | -.0331 | -.0385 | -.0057 | -.0192 |
| 1.0000 | -.0000 | -.037 | -.0021 |  .0033 |  .0009 | -.0005 | -.0003 |
| 5.0000 | -.0021 |       | -.0347 | -.0285 | -.0336 | -.0047 | -.0169 |
| 1.2500 | -.0001 | -.032 | -.0021 |  .0019 |  .0012 | -.0005 | -.0005 |

TABLE B.44. NUMERICAL VALUES FOR FACTORS IN EQUS. (3.51) TO (3.56).

SPR....... = .100
SPL....... = .006
PUL....... = 0.000
PÜR....... = 1.000
FDIS.R... = .300
K......... = .250

| KR KL | KWR KWL | KMC T | KQXR KQXL | KQYRN KQYLN | KQYRF KQYLF | KMXR KMXL | KMYR KMYL |
|---|---|---|---|---|---|---|---|
| 1.0000 | -.0002 | | -.0091 | -.0447 | -.0039 | -.0603 | -.0034 |
| .2500 | -.0000 | -.024 | -.0000 | .0111 | -.0000 | -.0000 | .0000 |
| 1.2000 | -.0004 | | -.0139 | -.0537 | -.0081 | -.0012 | -.0057 |
| .3000 | -.0000 | -.030 | -.0001 | .0133 | -.0000 | -.0000 | -.0000 |
| 1.4000 | -.0007 | | -.0178 | -.0588 | -.0129 | -.0021 | -.0081 |
| .3500 | -.0000 | -.035 | -.0002 | .0143 | -.0000 | -.0000 | -.0000 |
| 1.6000 | -.0010 | | -.0206 | -.0606 | -.0177 | -.0030 | -.0104 |
| .4000 | -.0000 | -.038 | -.0003 | .0145 | -.0000 | -.0001 | -.0000 |
| 1.8000 | -.0013 | | -.0223 | -.0603 | -.0221 | -.0038 | -.0124 |
| .4500 | -.0000 | -.040 | -.0005 | .0141 | -.0000 | -.0001 | -.0000 |
| 2.0000 | -.0015 | | -.0233 | -.0586 | -.0258 | -.0045 | -.0140 |
| .5000 | -.0000 | -.041 | -.0006 | .0133 | -.0001 | -.0001 | -.0000 |
| 3.0000 | -.0021 | | -.0230 | -.0449 | -.0346 | -.0057 | -.0175 |
| .7500 | -.0000 | -.039 | -.0014 | .0081 | -.0003 | -.0003 | -.0001 |
| 4.0000 | -.0021 | | -.0217 | -.0347 | -.0341 | -.0053 | -.0170 |
| 1.0000 | -.0000 | -.034 | -.0017 | .0045 | -.0007 | -.0004 | -.0003 |
| *.0000 | -.0019 | | -.0211 | -.0293 | -.0312 | -.0047 | -.0155 |
| 1.2500 | -.0001 | -.030 | -.0018 | .0027 | .0010 | -.0004 | -.0004 |

TABLE B.45. NUMERICAL VALUES FOR FACTORS IN EQUS. (3.51) TO (3.56).

```
SPR...... =  .200
SPL...... =  .013
PUL...... = 0.000
FOR...... = 1.000
POIS.R... =  .300
K........ =  .250
```

| RR / RL | KKR / KWL | KMC | KQXR / KQXL | KQYRN / KQYLN | KQYRF / KQYLF | KMXR / KMXL | KMYR / KMYL |
|---|---|---|---|---|---|---|---|
| 1.0000 | -.0001 |        | -.0055 | -.0269 |  .0024 |  .0010 | -.0033 |
|  .2500 | -.0000 | -.020  | -.0000 |  .0067 | -.0000 | -.0000 | -.0000 |
| 1.2000 | -.0003 |        | -.0087 | -.0341 | -.0051 | -.0009 | -.0052 |
|  .3000 | -.0000 | -.025  | -.0000 |  .0084 | -.0000 | -.0000 | -.0000 |
| 1.4000 | -.0005 |        | -.0117 | -.0391 | -.0065 | -.0004 | -.0072 |
|  .3500 | -.0000 | -.029  | -.0001 |  .0096 |  .0000 | -.0000 | -.0000 |
| 1.6000 | -.0007 |        | -.0141 | -.0422 | -.0122 | -.0002 | -.0090 |
|  .4000 | -.0000 | -.031  | -.0002 |  .0101 |  .0060 | -.0000 | -.0000 |
| 1.2000 | -.0009 |        | -.0158 | -.0437 | -.0157 | -.0009 | -.0106 |
|  .4500 | -.0000 | -.033  | -.0603 |  .0102 | -.0000 | -.0001 | -.0000 |
| 2.0000 | -.0011 |        | -.0169 | -.0441 | -.0189 |  .0016 | -.0120 |
|  .5000 | -.0000 | -.034  | -.0005 |  .0100 | -.0000 | -.0001 | -.0000 |
| 3.0000 | -.0017 |        | -.0179 | -.0388 | -.0282 | -.0036 | -.0152 |
|  .7500 | -.0000 | -.034  | -.0011 |  .0072 |  .0003 | -.0002 | -.0001 |
| 4.0000 | -.0018 |        | -.0171 | -.0322 | -.0297 | -.0044 | -.0151 |
| 1.0000 | -.0000 | -.031  | -.0014 |  .0045 |  .0006 | -.0003 | -.0003 |
| 5.0000 | -.0016 |        | -.0166 | -.0279 | -.0283 | -.0043 | -.0141 |
| 1.2500 | -.0000 | -.028  | -.0015 |  .0029 |  .0009 | -.0003 | -.0004 |

TABLE B.46. NUMERICAL VALUES FOR FACTORS IN EQUS. (3.51) TO (3.56).

307

SPR....... = .300
SPL....... = .019
POL....... = 0.000
POR....... = 1.000
POIS.F.... = .300
K......... = .250

| I | RR / RL | KVR / KVL | KMC | KQXR / KQXL | KQYRN / KQYLN | KQYRF / KQYLF | KMXR / KMXL | KMYR / KMYL | I |
|---|---------|-----------|-----|-------------|---------------|---------------|-------------|-------------|---|
| I | 1.0000 | -.0001 |      | -.0039 | -.0192 | -.0017 | .0016 | -.0032 | I |
| I | .2500  | -.0000 | -.018 | -.0000 | .0048 | .0000 | -.0000 | -.0000 | I |
| I | 1.2000 | -.0002 |      | -.0064 | -.0250 | -.0037 | .0018 | -.0050 | I |
| I | .3000  | -.0000 | -.023 | -.0000 | .0062 | .0000 | -.0000 | -.0000 | I |
| I | 1.4000 | -.0003 |      | -.0088 | -.0293 | -.0064 | .0017 | -.0067 | I |
| I | .3500  | -.0000 | -.026 | -.0001 | .0072 | .0000 | -.0000 | -.0000 | I |
| I | 1.5000 | -.0005 |      | -.0107 | -.0323 | -.0093 | .0013 | -.0083 | I |
| I | .4000  | -.0000 | -.028 | -.0002 | .0078 | .0000 | -.0000 | -.0000 | I |
| I | 1.8000 | -.0007 |      | -.0122 | -.0342 | -.0122 | .0007 | -.0097 | I |
| I | .4500  | -.0000 | -.030 | -.0003 | .0080 | .0000 | -.0000 | -.0000 | I |
| I | 2.0000 | -.0009 |      | -.0133 | -.0351 | -.0149 | .0001 | -.0108 | I |
| I | .5000  | -.0000 | -.031 | -.0064 | .0080 | .0000 | -.0001 | -.0000 | I |
| I | 3.0000 | -.0014 |      | -.0147 | -.0334 | -.0238 | -.0025 | -.0136 | I |
| I | .7500  | -.0000 | -.031 | -.0009 | .0062 | .0002 | .0002 | .0001 | I |
| I | 4.0000 | -.0016 |      | -.0143 | -.0292 | -.0261 | -.0036 | -.0136 | I |
| I | 1.0000 | -.0000 | -.029 | -.0012 | .0042 | .0005 | -.0003 | -.0002 | I |
| I | 5.0000 | -.0016 |      | -.0139 | -.0260 | -.0257 | -.0038 | -.0129 | I |
| I | 1.2500 | -.0000 | -.026 | -.0013 | .0026 | .0006 | -.0003 | -.0004 | I |

TABLE B.47. NUMERICAL VALUES FOR FACTORS IN EQUS. (3.51) TO (3.56).

SPR....... =   .400
SPL....... =   .025
POL....... = 0.000
PGR....... = 1.000
POIS.R.... =   .300
K......... =   .250

| KK RL | KWK KWL | KMC | KQXK KQXL | KQYRN KQYLN | KQYRF KQYLF | KMXR KMXL | KMYR KMYL |
|---|---|---|---|---|---|---|---|
| 1.0000 | -.0001 |        | -.0030 | -.0150 | -.0013 |  .0019 | -.0031 |
|  .2500 | -.0000 | -.017  | -.0000 |  .0037 | -.0000 |  .0000 | -.0000 |
| 1.2000 | -.0002 |        | -.0050 | -.0197 | -.0029 |  .0023 | -.0048 |
|  .3000 | -.0000 | -.021  | -.0000 |  .0049 | -.0000 | -.0000 | -.0000 |
| 1.4000 | -.0003 |        | -.0070 | -.0234 | -.0051 |  .0024 | -.0064 |
|  .3500 | -.0000 | -.024  | -.0001 |  .0057 | -.0000 | -.0000 | -.0000 |
| 1.6000 | -.0004 |        | -.0087 | -.0262 | -.0075 |  .0022 | -.0079 |
|  .4000 | -.0000 | -.026  | -.0001 |  .0063 | -.0000 | -.0000 | -.0000 |
| 1.8000 | -.0006 |        | -.0100 | -.0280 | -.0100 |  .0018 | -.0091 |
|  .4500 | -.0000 | -.027  | -.0002 |  .0066 | -.0000 | -.0000 | -.0000 |
| 2.0000 | -.0007 |        | -.0110 | -.0292 | -.0123 |  .0012 | -.0101 |
|  .5000 | -.0000 | -.028  | -.0003 |  .0067 | -.0000 | -.0000 | -.0000 |
| 3.0000 | -.0012 |        | -.0126 | -.0293 | -.0205 | -.0014 | -.0124 |
|  .7500 | -.0000 | -.028  | -.0008 |  .0055 |  .0002 | -.0001 | -.0001 |
| 4.0000 | -.0014 |        | -.0124 | -.0265 | -.0233 | -.0029 | -.0125 |
| 1.0000 | -.0000 | -.026  | -.0011 |  .0039 |  .0005 | -.0002 |  .0002 |
| 5.0000 | -.0015 |        | -.0121 | -.0241 | -.0234 | -.0034 | -.0119 |
| 1.2500 | -.0000 | -.024  | -.0012 |  .0027 |  .0007 | -.0003 | -.0003 |

TABLE B.48. NUMERICAL VALUES FOR FACTORS IN EQUS. (3.51) TO (3.56).

SPR........ = .500
SPL........ = .031
PUL........ = 0.0000
PQR........ = 1.000
POIS.R.... = .300
K.......... = .250

| RF RL | KWR KWL | KMC | KQXR KQXL | KQYRN KQYLN | KQYRF KQYLF | KMXR KMXL | KMYR KMYL |
|---|---|---|---|---|---|---|---|
| 1.0000 | -.0001 | | -.0025 | -.0123 | -.0011 | .0021 | -.0031 |
| .2500 | -.0000 | -.017 | -.0000 | .0031 | .0000 | .0000 | -.0000 |
| 1.2000 | -.0001 | | -.0041 | -.0162 | -.0024 | .0027 | -.0047 |
| .3000 | -.0000 | -.020 | -.0000 | .0040 | .0000 | .0000 | -.0000 |
| 1.4000 | -.0002 | | -.0058 | -.0195 | -.0042 | .0029 | -.0063 |
| .3500 | -.0000 | -.023 | -.0001 | .0048 | .0000 | -.0000 | -.0000 |
| 1.6000 | -.0004 | | -.0073 | -.0220 | -.0063 | .0028 | -.0076 |
| .4000 | -.0000 | -.024 | -.0001 | .0053 | .0000 | -.0000 | -.0000 |
| 1.8000 | -.0005 | | -.0084 | -.0237 | -.0084 | .0025 | -.0087 |
| .4500 | -.0000 | -.025 | -.0002 | .0056 | .0000 | -.0000 | -.0000 |
| 2.0000 | -.0006 | | -.0093 | -.0249 | -.0105 | .0020 | -.0096 |
| .5000 | -.0000 | -.026 | -.0003 | .0057 | .0000 | -.0000 | -.0000 |
| 3.0000 | -.0011 | | -.0110 | -.0259 | -.0161 | -.0007 | -.0115 |
| .7500 | -.0000 | -.026 | -.0007 | .0049 | .0002 | -.0001 | -.0001 |
| 4.0000 | -.0013 | | -.0109 | -.0242 | -.0210 | -.0023 | -.0115 |
| 1.0000 | -.0000 | -.024 | -.0010 | .0036 | .0004 | -.0002 | -.0002 |
| 5.0000 | -.0013 | | -.0107 | -.0224 | -.0214 | -.0030 | -.0110 |
| 1.2500 | -.0000 | -.022 | -.0011 | .0026 | .0006 | -.0002 | -.0003 |

TABLE B.49. NUMERICAL VALUES FOR FACTORS IN EQUS. (3.51) TO (3.56).

SPR...... = 0.000
SPL...... = 0.000
POL...... = 1.000
POR...... = 0.000
POIS.R... = .300
K........ = .250

| RK RL | KWR KWL | KMC | KQXR KQXL | KQYRN KQYLN | KQYRF KQYLF | KMXR KMXL | KMYR KMYL |
|---|---|---|---|---|---|---|---|
| 1.0000 | -.0009 | | -.0408 | -.1883 | -.0177 | -.0104 | -.0064 |
| .2500 | -.0000 | -.063 | -.0001 | .0469 | .0000 | -.0000 | -.0000 |
| 1.2000 | -.0020 | | -.0630 | -.2308 | -.0364 | -.0165 | -.0144 |
| .3000 | -.0000 | -.092 | -.0003 | .0570 | .0000 | -.0001 | -.0000 |
| 1.4000 | -.0035 | | -.0878 | -.2775 | -.0631 | -.0236 | -.0264 |
| .3500 | -.0000 | -.127 | -.0008 | .0677 | .0000 | -.0002 | -.0001 |
| 1.6000 | -.0035 | | -.1144 | -.3290 | -.0979 | -.0316 | -.0425 |
| .4000 | -.0000 | -.169 | -.0017 | .0788 | .0001 | -.0004 | -.0001 |
| 1.8000 | -.0040 | | -.1422 | -.3850 | -.1404 | -.0404 | -.0625 |
| .4500 | -.0000 | -.218 | -.0630 | .0901 | .0002 | -.0006 | -.0001 |
| 2.0000 | -.0110 | | -.1704 | -.4450 | -.1902 | -.0498 | -.0864 |
| .5000 | -.0000 | -.272 | -.0047 | .1013 | .0004 | -.0011 | -.0001 |
| 3.0000 | -.0305 | | -.2967 | -.7683 | -.5072 | -.1010 | -.2426 |
| .7500 | -.0363 | -.594 | -.0195 | .1472 | .0049 | -.0047 | -.0013 |
| 4.0000 | -.0510 | | -.3748 | -1.0536 | -.8343 | -.1471 | -.4071 |
| 1.0000 | -.0009 | -.908 | -.0370 | .1675 | .0160 | -.0094 | -.0056 |
| 5.0000 | -.0675 | | -.4121 | -1.2595 | -1.0938 | -.1812 | -.5390 |
| 1.2500 | -.0617 | -1.150 | -.0505 | .1683 | .0308 | -.0132 | -.0124 |

TABLE B.50. NUMERICAL VALUES FOR FACTORS IN EQUS. (3.51) TO (3.56).

311

```
SPR....... =   .100
SPL....... =   .006
PUL....... =  1.000
POR....... =  0.000
POIS.R.... =   .300
K......... =   .250
```

| KH / HL | KHR / KWL (T) | KMC | KQXR / KQXL | KQYRN / KQYLN | KQYRF / KQYLF | KMXR / KMXL | KMYR / KMYL |
|---|---|---|---|---|---|---|---|
| 1.0000 | -.0003 |  | -.0136 | -.0673 | -.0059 | -.0005 | -.0051 |
| .2500 | -.0000 | -.036 | -.0000 | .0168 | .0000 | -.0000 | .0000 |
| 1.2000 | -.0006 |  | -.0260 | -.1023 | -.0151 | -.0022 | -.0107 |
| .3000 | -.0000 | -.057 | -.0001 | .0253 | -.0000 | -.0000 | -.0000 |
| 1.4000 | -.0017 |  | -.0422 | -.1434 | -.0306 | -.0050 | -.0194 |
| .3500 | -.0000 | -.084 | -.0004 | .0350 | .0000 | -.0001 | .0000 |
| 1.6000 | -.0030 |  | -.0615 | -.1899 | .0534 | -.0091 | -.0314 |
| .4000 | -.0000 | -.118 | -.0009 | .0456 | .0000 | -.0002 | -.0000 |
| 1.8000 | -.0048 |  | -.0831 | -.2413 | -.0836 | -.0143 | -.0471 |
| .4500 | -.0000 | -.159 | -.0018 | .0567 | -.0001 | -.0003 | -.0000 |
| 2.0000 | -.0076 |  | -.1061 | -.2970 | -.1211 | -.0207 | -.0663 |
| .5000 | -.0003 | -.204 | -.0030 | .0679 | -.0003 | -.0006 | -.0000 |
| 3.0000 | -.0234 |  | -.2191 | -.6060 | -.3892 | -.0635 | -.2007 |
| .7500 | -.0002 | -.492 | -.0148 | .1173 | -.0037 | -.0033 | -.0013 |
| 4.0000 | -.0424 |  | -.2960 | -.8915 | -.6935 | -.1096 | -.3518 |
| 1.0000 | -.0007 | -.747 | -.0305 | .1437 | .0132 | -.0073 | -.0052 |
| 5.0000 | -.0584 |  | -.3352 | -1.1058 | -.9483 | -.1471 | -.4780 |
| 1.2500 | -.0014 | -1.023 | -.0431 | .1503 | .0264 | -.0109 | -.0111 |

TABLE B.51. NUMERICAL VALUES FOR FACTORS IN EQUS. (3.51) TO (3.56).

SPR........ = .200  
SPL........ = .013  
PGL........ = 1.000  
PGR........ = 0.000  
POIS.R.... = .300  
K.......... = .250  

| FK / FL | KWR / KWL | KMC | KQXR / KQXL | KQYRN / KQYLN | KQYRF / KQYLF | KMXR / KMXL | KMYR / KMYL |
|---|---|---|---|---|---|---|---|
| 1.0000 | -.0002 | | -.0081 | -.0405 | -.0035 | .0015 | -.0049 |
| .2500 | -.0000 | -.030 | -.0000 | .0101 | .0000 | -.0000 | -.0000 |
| 1.2000 | -.0005 | | -.0164 | -.0648 | -.0095 | .0016 | -.0097 |
| .3000 | -.0000 | -.048 | -.0001 | .0160 | .0000 | -.0000 | -.0000 |
| 1.4000 | -.0011 | | -.0278 | -.0950 | -.0202 | .0010 | -.0171 |
| .3500 | -.0000 | -.070 | -.0003 | .0232 | .0000 | -.0000 | -.0000 |
| 1.6000 | -.0021 | | -.0422 | -.1311 | -.0367 | -.0005 | -.0273 |
| .4000 | -.0000 | -.098 | -.0006 | .0315 | .0000 | -.0001 | -.0000 |
| 1.8000 | -.0034 | | -.0590 | -.1725 | -.0594 | -.0031 | -.0406 |
| .4500 | -.0000 | -.132 | -.0013 | .0405 | .0001 | -.0002 | -.0000 |
| 2.0000 | -.0051 | | -.0776 | -.2188 | -.0687 | -.0068 | -.0569 |
| .5000 | -.0000 | -.172 | -.0022 | .0501 | .0002 | -.0004 | -.0001 |
| 3.0000 | -.0149 | | -.1760 | -.4937 | -.3149 | -.0392 | -.1746 |
| .7500 | -.0002 | -.426 | -.0120 | .0958 | .0030 | -.0025 | -.0013 |
| 4.0000 | -.0361 | | -.2490 | -.7654 | -.5915 | -.0814 | -.3123 |
| 1.0000 | -.0006 | -.699 | -.0259 | .1240 | .0112 | -.0059 | -.0048 |
| 5.0000 | -.0514 | | -.2886 | -.9786 | -.8342 | -.1193 | -.4307 |
| 1.2500 | -.0013 | -.923 | -.0377 | .1340 | .0232 | -.0092 | -.0101 |

TABLE B.52. NUMERICAL VALUES FOR FACTORS IN EQUS. (3.51) TO (3.56).

SPK...... = .300
SPL...... = .019
PUL...... = 1.000
PQR...... = 0.000
PUIS.K... = .300
K........ = .250

| RK RL | KwK KwL | KMC | KQXR KQXL | KQYRN KQYLN | KQYRF KQYLF | KMXR KMXL | KMYR KMYL |
|---|---|---|---|---|---|---|---|
| 1.0000 | -.0001 |  | -.0058 | -.0289 | -.0025 | .0023 | -.0047 |
| .2500 | -.0000 | -.028 | -.0000 | .0072 | -.0000 | -.0000 | -.0000 |
| 1.2000 | -.0004 |  | -.0120 | -.0473 | -.0070 | .0034 | -.0093 |
| .3600 | -.0000 | -.043 | -.0001 | .0117 | .0000 | -.0000 | -.0000 |
| 1.4000 | -.0008 |  | -.0208 | -.0710 | -.0151 | .0040 | -.0160 |
| .3500 | -.0000 | -.063 | -.0002 | .0174 | .0000 | -.0000 | -.0000 |
| 1.6000 | -.0016 |  | -.0321 | -.1000 | -.0279 | .0040 | -.0252 |
| .4000 | -.0000 | -.088 | -.0005 | .0240 | -.0000 | .0001 | -.0000 |
| 1.8000 | -.0026 | -.118 | -.0457 | -.1341 | -.0461 | .0031 | -.0370 |
| .4500 | -.0000 |  | -.0010 | .0315 | -.0001 | -.0001 | -.0001 |
| 2.0000 | -.0040 | -.153 | -.0612 | -.1730 | -.0700 | .0013 | -.0516 |
| .5000 | -.0000 |  | -.0017 | .0396 | .0002 | -.0002 | -.0001 |
| 3.0000 | -.0159 | -.381 | -.1473 | -.4157 | -.2643 | -.0226 | -.1570 |
| .7500 | -.0001 |  | -.0101 | .0607 | .0025 | .0019 | -.0012 |
| 4.0000 | -.0315 | -.632 | -.2155 | -.6690 | -.5153 | -.0602 | -.2829 |
| 1.0000 | -.0005 |  | -.0225 | .1086 | .0098 | -.0048 | -.0045 |
| 5.0000 | -.0458 | -.843 | -.2544 | -.8755 | -.7438 | -.0970 | -.3936 |
| 1.2500 | -.0011 |  | -.0335 | .1203 | .0206 | -.0078 | -.0094 |

TABLE B.53. NUMERICAL VALUES FOR FACTORS IN EQUS. (3.51) TO (3.56).

SPR....... = .400
SPL....... = .025
PDL....... = 1.000
PUR....... = 0.000
PUIS.R... = .300
K......... = .250

| RR / HL | KWR / KWL | KMC | KQXR / KQXL | KQYRN / KQYLN | KQYRF / KQYLF | KMXR / KMXL | KMYR / KMYL |
|---|---|---|---|---|---|---|---|
| 1.0000 | -.0001 | | -.0045 | -.0225 | -.0020 | .0028 | -.0047 |
| .2500 | -.0000 | -.076 | -.0000 | .0056 | .0000 | .0000 | -.0000 |
| 1.2000 | -.0003 | | -.0094 | -.0373 | -.0055 | .0044 | -.0090 |
| .3000 | -.0003 | -.041 | -.0000 | .0092 | .0000 | -.0000 | -.0000 |
| 1.4000 | -.0007 | | -.0166 | -.0567 | -.0120 | .0058 | -.0154 |
| .3500 | -.0000 | -.059 | -.0002 | .0139 | .0000 | .0058 | -.0000 |
| 1.6000 | -.0013 | | -.0259 | -.0808 | -.0225 | .0068 | -.0239 |
| .4000 | -.0000 | -.081 | -.0004 | .0194 | .0000 | .0000 | -.0000 |
| 1.8000 | -.0021 | | -.0374 | -.1097 | -.0377 | .0071 | -.0348 |
| .4500 | -.0000 | -.109 | -.0008 | .0258 | .0000 | -.0001 | -.0001 |
| 2.0000 | -.0033 | | -.0505 | -.1431 | -.0578 | .0065 | -.0480 |
| .5000 | -.0000 | -.140 | -.0014 | .0328 | .0001 | -.0002 | -.0001 |
| 3.0000 | -.0137 | | -.1267 | -.3587 | -.2277 | -.0105 | -.1442 |
| .7500 | -.0001 | -.348 | -.0087 | .0697 | .0022 | -.0015 | -.0612 |
| 4.0000 | -.0279 | | -.1501 | -.5937 | -.4563 | -.0437 | -.2602 |
| 1.0000 | -.0005 | -.590 | -.0199 | .0965 | .0086 | -.0040 | -.0042 |
| 5.0000 | -.0413 | | -.2279 | -.7913 | -.6708 | -.0789 | -.3636 |
| 1.2500 | -.0010 | -.779 | -.0302 | .1090 | .0186 | -.0067 | -.0088 |

TABLE B.54. NUMERICAL VALUES FOR FACTORS IN EQUS. (3.51) TO (3.56).

SPR....... = .500
SPL....... = .031
POL....... = 1.000
POR....... = 0.000
POIS.R...= .300
K....... = .250

| RK / KL | KWK / KWL | KMC | KQXK / KQXL | KQYRN / KQYLN | KQYRF / KQYLF | KMXR / KMXL | KMYR / KMYL |
|---|---|---|---|---|---|---|---|
| 1.0000 | -.0001 |  | -.0037 | -.0184 | -.0016 | .0031 | -.0046 |
| .2500 | -.0000 | -.026 | -.0000 | .0046 | -.0000 | .0000 | -.0000 |
| 1.2000 | -.0002 |  | -.0678 | -.0308 | -.0045 | .0050 | -.0089 |
| .3000 | -.0000 | -.039 | -.0000 | .0076 | -.0000 | .0000 | -.0000 |
| 1.4000 | -.0006 |  | -.0138 | -.0472 | -.0100 | .0070 | -.0149 |
| .3500 | -.0000 | -.056 | -.0001 | .0115 | -.0000 | -.0000 | -.0000 |
| 1.6000 | -.0011 |  | -.0217 | -.0678 | -.0189 | .0086 | -.0230 |
| .4000 | -.0000 | -.077 | -.0003 | .0163 | -.0000 | -.0000 | -.0000 |
| 1.8000 | -.0016 |  | -.0316 | -.0928 | -.0318 | .0098 | -.0332 |
| .4500 | -.0000 | -.102 | -.0007 | .0218 | -.0000 | -.0600 | -.0001 |
| 2.0000 | -.0028 |  | -.0430 | -.1219 | -.0492 | .0102 | -.0456 |
| .5000 | -.0000 | -.131 | -.0012 | .0279 | -.0001 | -.0001 | -.0001 |
| 3.0000 | -.0120 |  | -.1112 | -.3154 | -.2000 | -.0014 | -.1345 |
| .7500 | -.0001 | -.324 | -.0676 | .0613 | .0019 | -.0011 | -.0012 |
| 4.0000 | -.0250 |  | -.1701 | -.5334 | -.4095 | -.0305 | -.2422 |
| 1.0000 | -.0004 | -.539 | -.0179 | .0866 | .0077 | -.0033 | -.0040 |
| 5.0000 | -.0376 |  | -.2065 | -.7215 | -.6108 | -.0639 | -.3391 |
| 1.2500 | -.0009 | -.726 | -.0274 | .0996 | .0169 | -.0058 | -.0063 |

TABLE B.55. NUMERICAL VALUES FOR FACTORS IN EQUS. (3.51) TO (3.56).

```
SPR...... = 0.000
SPL...... = C.000
PGL...... = 0.000
POR...... = 1.000
POIS.R... = .300
K........ = .500
```

| RR / RL | KWR / KWL | KMC | KQXR / KQXL | KQYRN / KQYLN | KQYRF / KQYLF | KMXR / KMXL | KMYR / KMYL |
|---|---|---|---|---|---|---|---|
| 1.0000 | -.0006 |       | -.0274 | -.1200 | -.0118 | -.0069 | -.0043 |
| .5000  | -.0000 | -.041 | -.0029 |  .0597 |  .0003 | -.0007 |  .0000 |
| 1.2000 | -.0010 |       | -.0339 | -.1134 | -.0194 | -.0088 | -.0077 |
| .6000  | -.0000 | -.048 | -.0048 |  .0560 |  .0007 | -.0011 | -.0001 |
| 1.4000 | -.0015 |       | -.0376 | -.1033 | -.0265 | -.0098 | -.0111 |
| .7000  | -.0001 | -.051 | -.0065 |  .0503 |  .0014 | -.0015 | -.0003 |
| 1.6000 | -.0018 |       | -.0394 | -.0925 | -.0325 | -.0103 | -.0142 |
| .8000  | -.0001 | -.053 | -.0079 |  .0440 |  .0023 | -.0019 | -.0007 |
| 1.8000 | -.0021 |       | -.0401 | -.0824 | -.0372 | -.0103 | -.0166 |
| .9000  | -.0002 | -.054 | -.0089 |  .0380 |  .0032 | -.0022 | -.0011 |
| 2.0000 | -.0023 |       | -.0401 | -.0736 | -.0406 | -.0101 | -.0186 |
| 1.0000 | -.0002 | -.053 | -.0097 |  .0327 |  .0041 | -.0024 | -.0015 |
| 3.0000 | -.0026 |       | -.0366 | -.0487 | -.0464 | -.0082 | -.0225 |
| 1.5000 | -.0064 | -.049 | -.0108 |  .0166 |  .0077 | -.0025 | -.0033 |
| 4.0000 | -.0028 |       | -.0380 | -.0419 | -.0451 | -.0069 | -.0224 |
| 2.0000 | -.0005 | -.045 | -.0104 |  .0114 |  .0095 | -.0022 | -.0044 |
| 5.0000 | -.0027 |       | -.0380 | -.0403 | -.0434 | -.0064 | -.0217 |
| 2.5000 | -.0006 | -.043 | -.0108 |  .0100 |  .0102 | -.0019 | -.0049 |

TABLE B.56. NUMERICAL VALUES FOR FACTORS IN EQUS. (3.51) TO (3.56).

317

```
SPR....... =  .100
SPL....... =  .025
PQL....... = 0.000
PUR.R..... = 1.000
POIS.R.... =  .300
K......... =  .500
```

| RR / RL | KKR / KKL | KMC | KQXR / KQXL | KQYRN / KQYLN | KQYRF / KQYLF | KMXR / KMXL | KMYR / KMYL |
|---|---|---|---|---|---|---|---|
| 1.0000 | -.0002 |       | -.0091 | -.0447 | -.0039 | -.0003 | -.0034 |
| .5000  | -.0000 | -.024 | -.0010 |  .0223 | -.0001 | -.0001 | -.0001 |
| 1.2000 | -.0004 |       | -.0139 | -.0538 | -.0081 | -.0012 | -.0057 |
| .6000  | -.0000 | -.030 | -.0020 |  .0266 | -.0003 | -.0003 | -.0002 |
| 1.4000 | -.0007 |       | -.0178 | -.0588 | -.0129 | -.0021 | -.0081 |
| .7000  | -.0000 | -.035 | -.0032 |  .0287 |  .0007 | -.0005 | -.0004 |
| 1.6000 | -.0010 |       | -.0206 | -.0607 | -.0177 | -.0030 | -.0104 |
| .8000  | -.0001 | -.038 | -.0043 |  .0291 |  .0012 | -.0008 | -.0007 |
| 1.8000 | -.0013 |       | -.0224 | -.0605 | -.0222 | -.0038 | -.0124 |
| .9000  | -.0001 | -.040 | -.0052 |  .0284 |  .0019 | -.0010 | -.0010 |
| 2.0000 | -.0015 |       | -.0234 | -.0591 | -.0260 | -.0045 | -.0141 |
| 1.0000 | -.0001 | -.041 | -.0060 |  .0269 |  .0026 | -.0012 | -.0013 |
| 3.0000 | -.0022 |       | -.0239 | -.0475 | -.0362 | -.0060 | -.0183 |
| 1.5000 | -.0003 | -.041 | -.0077 |  .0179 |  .0058 | -.0017 | -.0027 |
| 4.0000 | -.0024 |       | -.0234 | -.0402 | -.0383 | -.0060 | -.0191 |
| 2.0000 | -.0064 | -.039 | -.0078 |  .0124 |  .0077 | -.0018 | -.0037 |
| 5.0000 | -.0024 |       | -.0233 | -.0373 | -.0378 | -.0058 | -.0188 |
| 2.5000 | -.0005 | -.038 | -.0079 |  .0101 |  .0086 | -.0016 | -.0041 |

TABLE B.57. NUMERICAL VALUES FOR FACTORS IN EQUS. (3.51) TO (3.56).

SPR...... = .200
SPL...... = .050
POL...... = 0.000
PDR...... = 1.000
POIS.R... = .300
K........ = .500

| RR / RL | KWR / KWL | KMC | KQXR / KQXL | KQYRN / KQYLN | KQYRF / KQYLF | KMXR / KMXL | KMYR / KMYL |
|---|---|---|---|---|---|---|---|
| 1.0000 | -.0001 | -.020 | -.0055 | -.0269 | -.0024 | .0010 | -.0033 |
| .5000 | -.0000 | | -.0006 | .0134 | .0001 | -.0000 | -.0001 |
| 1.2000 | -.0003 | -.025 | -.0087 | -.0341 | -.0051 | -.0009 | -.0052 |
| .6000 | -.0000 | | -.0013 | .0169 | .0002 | -.0001 | -.0003 |
| 1.4000 | -.0005 | -.029 | -.0117 | -.0392 | -.0085 | .0004 | -.0072 |
| .7000 | -.0000 | | -.0021 | .0191 | .0004 | -.0002 | -.0004 |
| 1.6000 | -.0007 | -.031 | -.0141 | -.0423 | -.0122 | -.0002 | -.0090 |
| .8000 | -.0000 | | -.0029 | .0203 | .0008 | -.0003 | -.0007 |
| 1.8000 | -.0009 | -.033 | -.0158 | -.0438 | -.0158 | -.0009 | -.0107 |
| .9000 | -.0001 | | -.0037 | .0206 | .0013 | -.0005 | -.0009 |
| 2.0000 | -.0011 | -.035 | -.0170 | -.0443 | -.0190 | -.0016 | -.0120 |
| 1.0000 | -.0001 | | -.0044 | .0203 | .0019 | -.0006 | -.0012 |
| 3.0000 | -.0018 | -.036 | -.0184 | -.0403 | -.0292 | -.0040 | -.0157 |
| 1.5000 | -.0003 | | -.0061 | .0155 | .0047 | -.0012 | -.0024 |
| 4.0000 | -.0026 | -.034 | -.0182 | -.0359 | -.0325 | -.0048 | -.0165 |
| 2.0000 | -.0004 | | -.0063 | .0115 | .0064 | -.0014 | -.0032 |
| 5.0000 | -.0020 | -.033 | -.0181 | -.0335 | -.0330 | -.0050 | -.0165 |
| 2.5000 | -.0004 | | -.0064 | .0094 | .0074 | -.0014 | -.0036 |

TABLE B.58. NUMERICAL VALUES FOR FACTORS IN EQUS. (3.51) TO (3.56).

```
SPR...... = .300
SPL...... = .075
POL...... = 0.000
POR...... = 1.000
PGIS.R... = .300
K........ = .500
```

| RR / RL | KWR / KWL | KMC | KQXR / KQXL | KQYRN / KQYLN | KQYRF / KQYLF | KMXR / KMXL | KMYR / KMYL |
|---|---|---|---|---|---|---|---|
| 1.0000 | -.0001 |  | -.0039 | -.0192 | -.0017 | .0016 | -.0032 |
| .5000 | -.0000 | -.018 | -.0004 | -.0096 | -.0000 | .0000 | -.0001 |
| 1.2000 | -.0002 |  | -.0064 | -.0250 | -.0037 | .0018 | -.0050 |
| .8000 | -.0000 | -.023 | -.0009 | -.0123 | .0001 | .0000 | -.0003 |
| 1.4000 | -.0003 |  | -.0088 | -.0293 | -.0064 | .0017 | -.0067 |
| .7000 | -.0000 | -.026 | -.0016 | -.0143 | .0003 | -.0000 | -.0004 |
| 1.6000 | -.0005 |  | -.0107 | -.0323 | -.0093 | .0013 | -.0083 |
| .8000 | -.0000 | -.024 | -.0022 | -.0155 | .0006 | -.0001 | -.0007 |
| 1.8000 | -.0007 |  | -.0123 | -.0342 | -.0122 | -.0007 | -.0097 |
| .9000 | -.0001 | -.030 | -.0029 | -.0161 | .0010 | -.0002 | -.0009 |
| 2.0000 | -.0009 |  | -.0133 | -.0353 | -.0150 | -.0001 | -.0109 |
| 1.0000 | -.0001 | -.031 | -.0035 | -.0161 | .0015 | -.0003 | -.0011 |
| 3.0000 | -.0015 |  | -.0151 | -.0344 | -.0244 | -.0025 | -.0140 |
| 1.5000 | -.0002 | -.032 | -.0050 | -.0133 | .0039 | -.0008 | -.0022 |
| 4.0000 | -.0017 |  | -.0151 | -.0318 | -.0261 | -.0038 | -.0147 |
| 2.0000 | -.0003 | -.030 | -.0054 | -.0104 | .0055 | -.0011 | -.0028 |
| 5.0000 | -.0015 |  | -.0150 | -.0301 | -.0291 | -.0043 | -.0147 |
| 2.5000 | -.0004 | -.030 | -.0054 | .0086 | .0064 | -.0012 | -.0032 |

TABLE B.59. NUMERICAL VALUES FOR FACTORS IN EQUS. (3.51) TO (3.56).

SPR....... = .400
SPL....... = .100
PUL....... = 0.000
PGR....... = 1.000
POIS.R...= .300
K........= .500

| FR / RL | KWR / KWL | KMC | KQXR / KQXL | KQYRN / KQYLN | KQYRF / KQYLF | KMXR / KMXL | KMYR / KMYL |
|---|---|---|---|---|---|---|---|
| 1.0000 | -.0001 |  | -.0030 | -.0150 | -.0013 | .0019 | -.0031 |
| .5000 | -.0000 | -.017 | -.0003 | -.0075 | .0000 | .0001 | -.0001 |
| 1.2000 | -.0002 |  | -.0050 | -.0197 | -.0029 | .0023 | -.0046 |
| .6000 | -.0000 | -.021 | -.0007 | .0097 | .0001 | .0001 | -.0003 |
| 1.4000 | -.0003 |  | -.0070 | -.0234 | -.0051 | .0024 | -.0064 |
| .7000 | -.0000 | -.024 | -.0012 | .0114 | .0003 | .0001 | -.0005 |
| 1.6000 | -.0004 |  | -.0087 | -.0262 | -.0075 | .0022 | -.0079 |
| .8000 | -.0000 | -.026 | -.0018 | -.0126 | .0005 | .0001 | -.0007 |
| 1.8000 | -.0006 |  | -.0100 | -.0281 | -.0100 | .0018 | -.0091 |
| .9000 | -.0000 | -.027 | -.0024 | .0132 | .0008 | .0000 | -.0009 |
| 2.0000 | -.0007 |  | -.0110 | -.0292 | -.0124 | .0012 | -.0101 |
| 1.0000 | -.0001 | -.028 | -.0029 | .0134 | .0012 | -.0001 | -.0011 |
| 3.0000 | -.0013 |  | -.0128 | -.0299 | -.0210 | .0015 | .0127 |
| 1.5000 | -.0002 | -.029 | -.0043 | .0116 | .0033 | -.0006 | -.0020 |
| 4.0000 | -.0015 |  | -.0129 | -.0283 | -.0247 | -.0030 | -.0133 |
| 2.0000 | -.0003 | -.028 | -.0046 | .0093 | .0048 | -.0009 | -.0026 |
| 5.0000 | -.0016 |  | -.0129 | -.0272 | -.0259 | -.0037 | -.0133 |
| 2.5000 | -.0003 | -.027 | -.0047 | .0079 | .0057 | -.0010 | -.0029 |

TABLE B.60. NUMERICAL VALUES FOR FACTORS IN EQUS. (3.51) TO (3.56).

```
SPR........ =  .500
SPL........ =  .125
PGL........ = 0.000
PGR........ = 1.600
POIS.R..... =  .300
K.......... =  .500
```

| I KK RL I | I KwR KwL I | I KMC I | I KQXR KQXL I | I KQYRN KQYLN I | I KQYRF KQYLF I | I KMXR KMXL I | I KMYR KMYL I |
|---|---|---|---|---|---|---|---|
| 1.0000 | -.0001 | | -.0025 | -.0123 | -.0011 | .0021 | -.0031 |
| .5000 | -.0006 | -.017 | -.0003 | .0061 | .0000 | .0001 | -.0001 |
| 1.2000 | -.0001 | | -.0041 | -.0162 | -.0024 | .0027 | -.0047 |
| .6000 | -.0000 | -.020 | -.0006 | .0080 | .0001 | .0001 | -.0003 |
| 1.4000 | -.0002 | | -.0058 | -.0195 | -.0042 | .0029 | -.0063 |
| .7000 | -.0000 | -.023 | -.0010 | .0095 | .0002 | .0002 | -.0005 |
| 1.6000 | -.0004 | | -.0073 | -.0220 | -.0063 | .0028 | -.0076 |
| .8000 | -.0000 | -.024 | -.0015 | .0106 | .0004 | .0001 | -.0087 |
| 1.8000 | -.0005 | | -.0084 | -.0238 | -.0084 | .0025 | -.0008 |
| .9000 | -.0000 | -.026 | -.0020 | .0112 | -.0007 | .0001 | -.0096 |
| 2.0000 | -.0006 | | -.0093 | -.0250 | -.0105 | .0020 | -.0010 |
| 1.0000 | -.0001 | -.026 | -.0624 | .0114 | .0011 | .0000 | -.0117 |
| 3.0000 | -.0011 | | -.0111 | -.0264 | -.0184 | -.0007 | -.0019 |
| 1.5000 | -.0002 | -.026 | -.0037 | .0103 | .0029 | -.0004 | -.0122 |
| 4.0000 | -.0012 | | -.0113 | -.0255 | -.0220 | -.0024 | -.0024 |
| 2.0000 | -.0002 | -.025 | -.0041 | .0085 | .0043 | -.0007 | -.0121 |
| 5.0000 | -.0014 | | -.0113 | -.0247 | -.0234 | -.0032 | -.0027 |
| 2.5000 | -.0003 | -.024 | -.0042 | .0072 | .0051 | -.0008 | |

TABLE B.61. NUMERICAL VALUES FOR FACTORS IN EQS. (3.51) TO (3.56).

SPK........ = 0.000
SPL........ = 0.000
POL........ = 1.000
PUK........ = 0.000
POIS.R.... = .300
K.......... = .500

| RR / KL | KWR / KWL | KMC | KQXR / KQXL | KQYRN / KQYLN | KQYRF / KQYLF | KMXR / KMXL | KMYR / KMYL |
|---|---|---|---|---|---|---|---|
| 1.0000 | -.0009 | | -.0399 | -.1827 | -.0172 | -.0101 | -.0063 |
| .5000 | -.0000 | -.061 | -.0043 | -.0910 | -.0004 | -.0010 | -.0001 |
| 1.2000 | -.0016 | | -.0591 | -.2144 | -.0341 | -.0154 | -.0135 |
| .6000 | -.0001 | -.086 | -.0084 | -.1060 | .0012 | -.0019 | -.0002 |
| 1.4000 | -.0031 | | -.0777 | -.2420 | -.0557 | -.0208 | -.0233 |
| .7000 | -.0002 | -.112 | -.0137 | -.1181 | .0029 | -.0032 | -.0007 |
| 1.6000 | -.0045 | | -.0944 | -.2649 | -.0803 | -.0259 | -.0349 |
| .8000 | -.0003 | -.138 | -.0194 | -.1269 | .0055 | -.0047 | -.0017 |
| 1.8000 | -.0060 | | -.1084 | -.2835 | -.1061 | -.0304 | -.0473 |
| .9000 | -.0005 | -.163 | -.0252 | -.1327 | .0090 | -.0063 | -.0031 |
| 2.0000 | -.0076 | | -.1197 | -.2982 | -.1316 | -.0343 | -.0599 |
| 1.0000 | -.0007 | -.166 | -.0306 | -.1359 | .0132 | -.0077 | -.0048 |
| 3.0000 | -.0141 | | -.1467 | -.3326 | -.2341 | -.0456 | -.1123 |
| 1.5000 | -.0021 | -.268 | -.0483 | -.1290 | .0373 | -.0128 | -.0160 |
| 4.0000 | -.0177 | | -.1520 | -.3382 | -.2691 | -.0493 | -.1417 |
| 2.0000 | -.0033 | -.307 | -.0541 | -.1125 | .0566 | -.0144 | -.0258 |
| 5.0000 | -.0195 | | -.1528 | -.3368 | -.3149 | -.0502 | -.1559 |
| 2.5000 | -.0041 | -.323 | -.0557 | .0996 | .0688 | -.0144 | -.0325 |

TABLE B.62. NUMERICAL VALUES FOR FACTORS IN EQUS. (3.51) TO (3.56).

323

```
SPR........= .100
SPL........= .025
POL........= 1.000
POR........= 0.000
POIS.R.....= .360
K..........= .500
```

| RK / RL | KWR / KWL | KMC | KQXR / KQXL | KQYRN / KQYLN | KQYRF / KQYLF | KMXR / KMXL | KMYR / KMYL |
|---|---|---|---|---|---|---|---|
| 1.0000 | -.0003 |  | -.0132 | -.0656 | -.0057 | -.0005 | -.0050 |
| .5000 | -.0000 | -.035 | -.0014 | .0327 | .0001 | -.0002 | -.0001 |
| 1.2000 | -.0008 |  | -.0243 | -.0957 | -.0141 | -.0020 | -.0160 |
| .6000 | -.0000 | -.053 | -.0035 | .0473 | .0005 | -.0005 | -.0004 |
| 1.4000 | -.0015 |  | -.0372 | -.1262 | -.0270 | -.0044 | -.0171 |
| .7000 | -.0001 | -.074 | -.0066 | .0617 | .0014 | -.0011 | -.0009 |
| 1.6000 | -.0025 |  | -.0504 | -.1549 | -.0437 | -.0074 | -.0258 |
| .8000 | -.0002 | -.097 | -.0105 | .0745 | .0030 | -.0019 | -.0016 |
| 1.8000 | -.0036 |  | -.0628 | -.1807 | -.0630 | -.0108 | -.0355 |
| .9000 | -.0003 | -.119 | -.0149 | .0850 | .0053 | -.0028 | -.0027 |
| 2.0000 | -.0048 |  | -.0736 | -.2029 | -.0836 | -.0143 | -.0457 |
| 1.0000 | -.0004 | -.140 | -.0192 | .0931 | .0083 | -.0039 | -.0041 |
| 3.0000 | -.0107 |  | -.1030 | -.2678 | -.1775 | -.0290 | -.0912 |
| 1.5000 | -.0016 | -.220 | -.0355 | .1059 | -.0280 | -.0085 | -.0132 |
| 4.0000 | -.0144 |  | -.1100 | -.2082 | -.2356 | -.0372 | -.1190 |
| 2.0000 | -.0026 | -.260 | -.0417 | .0986 | .6455 | -.0108 | -.0216 |
| 5.0000 | -.0164 |  | -.1112 | -.2926 | -.2656 | -.0410 | -.1333 |
| 2.5000 | -.0034 | -.278 | -.0434 | .0891 | .0572 | -.0116 | -.0276 |

TABLE B.63. NUMERICAL VALUES FOR FACTORS IN EQUS. (3.51) TO (3.56).

SPR....... = .200
SPL....... = .050
PGL....... = 1.000
PQR....... = 0.000
POIS.R... = .300
K......... = .500

| KR / KL | KWR / KWL | KMC | KQXR / KQXL | KQYRN / KQYLN | KQYRF / KQYLF | KMXR / KMXL | KMYR / KMYL |
|---|---|---|---|---|---|---|---|
| 1.0000 | -.0002 | | -.0079 | -.0395 | -.0035 | .0015 | -.0047 |
| .5000 | -.0000 | -.029 | -.0009 | -.0197 | -.0001 | -.0000 | -.0002 |
| 1.2000 | -.0005 | | -.0153 | -.0606 | -.0089 | .0015 | -.0091 |
| .6000 | -.0000 | -.044 | -.0022 | -.0300 | -.0003 | -.0001 | -.0004 |
| 1.4000 | -.0010 | | -.0246 | -.0837 | -.0176 | .0009 | -.0151 |
| .7000 | -.0001 | -.062 | -.0044 | -.0409 | -.0009 | -.0004 | -.0009 |
| 1.6000 | -.0017 | | -.0346 | -.1070 | -.0300 | -.0004 | -.0224 |
| .8000 | -.0001 | -.080 | -.0072 | -.0514 | -.0021 | -.0008 | -.0016 |
| 1.8000 | -.0026 | | -.0445 | -.1292 | -.0448 | -.0024 | -.0306 |
| .9000 | -.0002 | -.099 | -.0105 | -.0608 | -.0038 | -.0014 | -.0026 |
| 2.0000 | -.0035 | | -.0536 | -.1496 | -.0611 | -.0047 | -.0392 |
| 1.0000 | -.0003 | -.117 | -.0141 | -.0687 | -.0061 | -.0021 | -.0038 |
| 3.0000 | -.0095 | | -.0613 | -.2177 | -.1422 | -.0180 | -.0784 |
| 1.5000 | -.0121 | -.189 | -.0283 | -.0865 | -.0224 | -.0058 | -.0115 |
| 4.0000 | -.0022 | | -.0892 | -.2454 | -.1974 | -.0277 | -.1033 |
| 2.6000 | -.0022 | -.226 | -.0344 | -.0646 | -.0379 | -.0082 | -.0189 |
| 5.0000 | -.0141 | | -.0908 | -.2549 | -.2280 | -.0333 | -.1165 |
| 2.5000 | -.0029 | -.244 | -.0362 | -.0787 | -.0488 | -.0094 | -.0241 |

**TABLE B.64. NUMERICAL VALUES FOR FACTORS IN EQUS. (3.51) TO (3.56).**

SPR...... = .300
SPL...... = .075
PGL...... = 1.000
POR...... = 0.000
POIS.R... = .300
K........ = .500

| RR KL | KWR KWL | KMC | KQXR KQXL | KQYRN KQYLN | KQYRF KQYLF | KMXR KMXL | KMYR KMYL |
|---|---|---|---|---|---|---|---|
| 1.0000 | -.0001 | | -.0057 | -.0282 | -.0025 | .0023 | -.0046 |
| .5000 | -.0000 | -.027 | -.0006 | .0140 | .0001 | .0001 | -.0002 |
| 1.2000 | -.0004 | | -.0112 | -.0443 | -.0065 | .0032 | -.0067 |
| .6000 | -.0000 | -.040 | -.0016 | .0219 | .0002 | .0001 | -.0005 |
| 1.4000 | -.0007 | | -.0183 | -.0625 | -.0133 | .0035 | -.0141 |
| .7000 | -.0000 | -.055 | -.0033 | .0306 | .0007 | -.0000 | -.0009 |
| 1.6000 | -.0013 | | -.0263 | -.0016 | -.0229 | -.0033 | -.0206 |
| .8000 | -.0001 | -.072 | -.0055 | .0392 | .0016 | .0002 | -.0016 |
| 1.8000 | -.0020 | | -.0345 | -.1005 | -.0347 | -.0023 | -.0278 |
| .9000 | -.0002 | -.088 | -.0082 | .0473 | .0029 | -.0006 | .0025 |
| 2.0000 | -.0026 | | -.0422 | -.1183 | -.0482 | .0608 | .0354 |
| 1.0000 | -.0063 | -.104 | -.0111 | .0543 | .0048 | -.0010 | -.0036 |
| 3.0000 | -.0071 | | -.0674 | -.1826 | -.1186 | -.0105 | -.0699 |
| 1.5000 | -.0010 | -.167 | -.0235 | .0727 | .0186 | -.0040 | -.0104 |
| 4.0000 | -.0104 | | -.0755 | -.2126 | -.1695 | -.0206 | -.0919 |
| 2.0000 | -.0019 | -.201 | -.0293 | .0738 | .0325 | -.0063 | -.0206 |
| 5.0000 | -.0123 | | -.0773 | -.2247 | -.1993 | -.0271 | -.1039 |
| 2.5000 | -.0025 | -.218 | -.0312 | .0699 | .0425 | -.0077 | -.0215 |

**TABLE B.65.** NUMERICAL VALUES FOR FACTORS IN EQUS. (3.51) TO (3.56).

326

SPR........= .400
SFL........= .100
PDL........= 1.000
PDR........= 0.000
PDIS.R.....= .300
K..........= .500

| RR / RL | KYR / KWL | KMC | KQXR / KQXL | KQYKN / KQYLN | KQYRF / KQYLF | KMXR / KMXL | KMYR / KMYL |
|---|---|---|---|---|---|---|---|
| 1.0000 | -.0001 | -.026 | -.0044 | -.0219 | -.0019 | .0027 | -.0046 |
| .5000 | -.0000 | | -.0005 | .0109 | .0000 | .0001 | -.0002 |
| 1.2000 | -.0003 | -.038 | -.0088 | -.0349 | -.0051 | .0041 | -.0085 |
| .6000 | -.0000 | | -.0013 | .0173 | .0002 | .0002 | -.0005 |
| 1.4000 | -.0006 | -.052 | -.0146 | -.0499 | -.0106 | .0051 | -.0135 |
| .7000 | -.0000 | | -.0026 | .0244 | .0006 | .0002 | -.0009 |
| 1.6600 | -.0010 | -.066 | -.0213 | -.0660 | -.0185 | .0055 | -.0195 |
| .8000 | -.0001 | | -.0044 | .0317 | .0013 | .0001 | -.0016 |
| 1.8000 | -.0016 | -.081 | -.0282 | -.0821 | -.0283 | .0053 | -.0261 |
| .9000 | -.0001 | | -.0067 | .0387 | .0024 | -.0000 | -.0024 |
| 2.0000 | -.0023 | -.095 | -.0348 | -.0977 | -.0398 | .0044 | -.0330 |
| 1.0000 | -.0002 | | -.0091 | .0449 | .0040 | -.0003 | -.0034 |
| 3.0000 | -.0061 | -.152 | -.0576 | -.1571 | -.1016 | -.0052 | -.0638 |
| 1.5000 | -.0009 | | -.0201 | .0626 | .0160 | -.0027 | -.0097 |
| 4.0000 | -.0091 | -.183 | -.0655 | -.1872 | -.1485 | -.0152 | -.0834 |
| 2.0000 | -.0016 | | -.0255 | .0652 | .0284 | -.0049 | -.0154 |
| 5.0000 | -.0109 | -.197 | -.0675 | -.2005 | -.1769 | -.0222 | -.0941 |
| 2.5000 | -.0022 | | -.0274 | .0626 | .0376 | -.0064 | -.0195 |

TABLE B.66. NUMERICAL VALUES FOR FACTORS IN EQUS. (3.51) TO (3.56).

SPR........ = .500
SPL........ = .125
PGL........ = 1.000
PQF........ = 0.000
PUIS.R.... = .300
K.......... = .500

| RR KL | KWR KWL | KMC | KQXR KQXL | KQYRN KQYLN | KQYRF KQYLF | KMXR KMXL | KMYR KMYL |
|---|---|---|---|---|---|---|---|
| 1.0000 | -.0001 |  | -.0036 | -.0180 | -.0016 | .0030 | -.0045 |
| .5000 | -.0000 | -.025 | -.0004 | -.0089 | .0000 | .0001 | -.0002 |
| 1.2000 | -.0002 |  | -.0073 | -.0288 | -.0042 | .0047 | -.0083 |
| .6000 | -.0000 | -.036 | -.0010 | .0142 | .0002 | .0002 | -.0005 |
| 1.4000 | -.0005 |  | -.0122 | -.0415 | -.0088 | .0061 | -.0132 |
| .7000 | -.0000 | -.049 | -.0022 | .0203 | .0005 | .0003 | -.0010 |
| 1.6000 | -.0009 |  | -.0178 | -.0553 | -.0155 | .0071 | -.0188 |
| .8000 | -.0001 | -.063 | -.0037 | .0266 | .0011 | .0004 | -.0016 |
| 1.8000 | -.0014 |  | -.0238 | -.0695 | -.0240 | .0073 | -.0249 |
| .9000 | -.0001 | -.076 | -.0056 | .0327 | .0020 | .0003 | -.0024 |
| 2.0000 | -.0020 |  | -.0296 | -.0833 | -.0338 | .0069 | -.0312 |
| 1.0000 | -.0002 | -.089 | -.0078 | .0383 | .0034 | .0001 | -.0034 |
| 3.0000 | -.0053 |  | -.0503 | -.1378 | -.0489 | -.0011 | -.0592 |
| 1.5000 | -.0008 | -.140 | -.0176 | .0549 | .0140 | -.0017 | -.0091 |
| 4.0000 | -.0061 |  | -.0579 | -.1670 | -.1321 | -.0110 | -.0768 |
| 2.0000 | -.0015 | -.169 | -.0226 | .0583 | .0253 | -.0038 | -.0143 |
| 5.0000 | -.0098 |  | -.0600 | -.1808 | -.1589 | -.0183 | -.0862 |
| 2.5000 | -.0020 | -.181 | -.0245 | .0566 | .0338 | -.0053 | -.0180 |

TABLE B.67. NUMERICAL VALUES FOR FACTORS IN EQUS. (3.51) TO (3.56).

```
SPR....... =  0.000
SPL....... =  0.000
PGL....... =  0.000
POR....... =  1.000
POIS.R.... =   .300
K......... =   .750
```

| RR / RL | KWR / KWL | KMC | KQXR / KQXL | KQYRN / KQYLN | KQYRF / KQYLF | KMXR / KMXL | KMYR / KMYL |
|---|---|---|---|---|---|---|---|
| 1.0000 | -.0006 |      | -.0275 | -.1202 | -.0116 | -.0069 | -.0043 |
|  .7500 | -.0002 | -.042 | -.0125 |  .0898 |  .0031 | -.0030 | -.0009 |
| 1.2000 | -.0011 |      | -.0341 | -.1141 | -.0195 | -.0088 | -.0076 |
|  .9000 | -.0003 | -.048 | -.0170 |  .0847 |  .0061 | -.0042 | -.0021 |
| 1.4000 | -.0015 |      | -.0381 | -.1049 | -.0269 | -.0099 | -.0113 |
| 1.0500 | -.0005 | -.052 | -.0204 |  .0770 |  .0095 | -.0051 | -.0036 |
| 1.6000 | -.0019 |      | -.0403 | -.0951 | -.0332 | -.0105 | -.0145 |
| 1.2000 | -.0007 | -.055 | -.0226 |  .0687 |  .0128 | -.0057 | -.0051 |
| 1.8000 | -.0022 |      | -.0413 | -.0862 | -.0384 | -.0107 | -.0172 |
| 1.3500 | -.0009 | -.056 | -.0239 |  .0611 |  .0159 | -.0061 | -.0066 |
| 2.0000 | -.0025 |      | -.0418 | -.0785 | -.0426 | -.0107 | -.0195 |
| 1.5000 | -.0010 | -.056 | -.0247 |  .0544 |  .0186 | -.0063 | -.0080 |
| 3.0000 | -.0032 |      | -.0418 | -.0580 | -.0522 | -.0094 | -.0253 |
| 2.2500 | -.0016 | -.055 | -.0257 |  .0357 |  .0266 | -.0058 | -.0125 |
| 4.0000 | -.0033 |      | -.0417 | -.0531 | -.0538 | -.0085 | -.0267 |
| 3.0000 | -.0018 | -.054 | -.0257 |  .0306 |  .0293 | -.0051 | -.0142 |
| 5.0000 | -.0034 |      | -.0417 | -.0522 | -.0538 | -.0081 | -.0269 |
| 3.7500 | -.0018 | -.054 | -.0257 |  .0293 |  .0300 | -.0047 | -.0148 |

TABLE B.68. NUMERICAL VALUES FOR FACTORS IN EQUS. (3.51) TO (3.56).

SPR....... = .100
SPL....... = .056
PUL....... = 0.000
PQR....... = 1.000
PUIS.R... = .300
K......... = .750

| KR / RL | KWR / KWL | KMC | KQXR / KQXL | KQYRN / KQYLN | KQYRF / KQYLF | KMXR / KMXL | KMYR / KMYL |
|---|---|---|---|---|---|---|---|
| 1.0000 / .7500 | -.0002 / -.0001 | -.024 | -.0091 / -.0042 | -.0447 / .0334 | -.0039 / .0010 | -.0003 / -.0003 | -.0034 / -.0010 |
| 1.2000 / .9000 | -.0004 / -.0001 | -.030 | -.0139 / -.0070 | -.0538 / .0400 | -.0081 / .0025 | -.0012 / -.0008 | -.0057 / -.0018 |
| 1.4000 / 1.0500 | -.0007 / -.0002 | -.035 | -.0179 / -.0097 | -.0591 / .0435 | -.0130 / .0046 | -.0021 / -.0014 | -.0082 / -.0028 |
| 1.6000 / 1.2000 | -.0010 / -.0004 | -.038 | -.0208 / -.0119 | -.0613 / .0446 | -.0179 / .0069 | -.0031 / -.0019 | -.0105 / -.0039 |
| 1.8000 / 1.3500 | -.0013 / -.0005 | -.041 | -.0228 / -.0135 | -.0616 / .0441 | -.0225 / .0093 | -.0039 / -.0024 | -.0126 / -.0050 |
| 2.0000 / 1.5000 | -.0015 / -.0006 | -.043 | -.0240 / -.0147 | -.0607 / .0427 | -.0266 / .0115 | -.0046 / -.0029 | -.0144 / -.0061 |
| 3.0000 / 2.2500 | -.0023 / -.0011 | -.045 | -.0254 / -.0165 | -.0518 / .0332 | -.0389 / .0195 | -.0064 / -.0039 | -.0197 / -.0097 |
| 4.0000 / 3.0000 | -.0026 / -.0014 | -.044 | -.0253 / -.0166 | -.0463 / .0275 | -.0429 / .0230 | -.0067 / -.0040 | -.0214 / -.0113 |
| 5.0000 / 3.7500 | -.0027 / -.0015 | -.044 | -.0253 / -.0166 | -.0443 / .0254 | -.0438 / .0242 | -.0067 / -.0039 | -.0218 / -.0120 |

TABLE B.69. NUMERICAL VALUES FOR FACTORS IN EQUS. (3.51) TO (3.56).

SPR........= .200
SPL........= .113
PDL........= 0.000
PDR........= 1.000
PDIS.R...= .300
K..........= .750

| KK / KL | KWK / KVL | KMC | KQXR / KQXL | KQYRN / KQYLN | KQYRF / KQYLF | KMXR / KMXL | KMYR / KMYL |
|---|---|---|---|---|---|---|---|
| 1.0000 | -.0001 |  | -.0055 | -.0269 | -.0024 | .0010 | -.0033 |
| .7500 | -.0000 | -.020 | -.0025 | .0201 | .0006 | .0002 | -.0010 |
| 1.2000 | -.0003 |  | -.0068 | -.0341 | -.0051 | .0009 | -.0052 |
| .9000 | -.0001 | -.025 | -.0044 | .0254 | .0016 | .0001 | -.0017 |
| 1.4000 | -.0005 |  | -.0118 | -.0393 | -.0085 | .0004 | -.0072 |
| 1.0500 | -.0002 | -.029 | -.0064 | .0289 | .0030 | -.0002 | -.0026 |
| 1.6000 | -.0007 |  | -.0142 | -.0425 | -.0123 | -.0002 | -.0091 |
| 1.2000 | -.0003 | -.032 | -.0081 | .0309 | -.0047 | -.0005 | -.0035 |
| 1.8000 | -.0009 |  | -.0160 | -.0442 | -.0159 | -.0009 | -.0108 |
| 1.3500 | -.0004 | -.034 | -.0095 | .0317 | .0065 | -.0009 | -.0044 |
| 2.0000 | -.0011 |  | -.0172 | -.0450 | -.0193 | -.0016 | -.0122 |
| 1.5000 | -.0005 | -.035 | -.0106 | .0317 | .0083 | -.0012 | -.0052 |
| 3.0000 | -.0018 |  | -.0192 | -.0426 | -.0306 | -.0041 | -.0165 |
| 2.2500 | -.0009 | -.036 | -.0126 | .0275 | .0153 | -.0025 | -.0081 |
| 4.0000 | -.0022 |  | -.0193 | -.0394 | -.0351 | -.0052 | -.0180 |
| 3.0000 | -.0011 | -.037 | -.0128 | .0237 | .0167 | -.0030 | -.0095 |
| 5.0000 | -.0023 |  | -.0193 | -.0378 | -.0366 | -.0055 | -.0184 |
| 3.7500 | -.0012 | -.037 | -.0128 | .0218 | .0201 | -.0032 | -.0101 |

TABLE B.70. NUMERICAL VALUES FOR FACTORS IN EQUS. (3.51) TO (3.56).

331

SPR...... = .300
SPL...... = .169
PUL...... = 0.000
POR...... = 1.000
POIS.R... = .300
K........ = .750

| KR / KL | KWR / KWL | KMC | KQXR / KQXL | KQYRN / KQYLN | KQYKF / KQYLF | KMXR / KMXL | KMYR / KMYL |
|---|---|---|---|---|---|---|---|
| 1.0000 | -.0001 | | -.0039 | -.0192 | -.0017 | .0016 | -.0032 |
| .7500 | -.0000 | -.018 | -.0018 | .0144 | .0004 | .0004 | -.0010 |
| 1.2000 | -.0002 | | -.0064 | -.0250 | -.0037 | .0018 | -.0050 |
| .9000 | -.0001 | -.023 | -.0032 | .0186 | .0012 | .0005 | -.0017 |
| 1.4000 | -.0004 | | -.0088 | -.0294 | -.0064 | .0017 | -.0067 |
| 1.0500 | -.0001 | -.026 | -.0048 | .0216 | .0022 | .0004 | -.0025 |
| 1.6000 | -.0005 | | -.0108 | -.0324 | -.0093 | .0013 | -.0083 |
| 1.2000 | -.0002 | -.028 | -.0062 | .0236 | .0036 | .0003 | -.0033 |
| 1.8000 | -.0007 | | -.0123 | -.0344 | -.0123 | .0007 | -.0098 |
| 1.3500 | -.0003 | -.030 | -.0073 | .0247 | .0050 | -.0000 | -.0041 |
| 2.0000 | -.0009 | | -.0135 | -.0356 | -.0151 | -.0001 | -.0110 |
| 1.5000 | -.0004 | -.031 | -.0083 | .0251 | .0065 | -.0003 | -.0047 |
| 3.0000 | -.0015 | | -.0155 | -.0357 | -.0252 | -.0026 | -.0144 |
| 2.2500 | -.0007 | -.033 | -.0102 | .0232 | .0126 | -.0040 | -.0071 |
| 4.0000 | -.0018 | | -.0157 | -.0339 | -.0296 | -.0023 | -.0155 |
| 3.0000 | -.0004 | -.032 | -.0105 | .0205 | .0158 | -.0046 | -.0082 |
| 5.0000 | -.0019 | | -.0157 | -.0327 | -.0313 | -.0026 | -.0159 |
| 3.7500 | -.0011 | -.032 | -.0106 | .0190 | .0172 | | -.0087 |

TABLE B.71. NUMERICAL VALUES FOR FACTORS IN EQUS. (3.51) TO (3.56).

```
SPR....... =  .400
SPL....... =  .225
PQL....... = 0.000
PQR....... = 1.000
POIS.K... =  .300
K......... =  .750
```

| RH | RL | KWR | KWL | KMC | KQXR | KQXL | KQYRN | KQYLN | KQYRF | KQYLF | KMXR | KMXL | KMYR | KMYL |
|---|---|---|---|---|---|---|---|---|---|---|---|---|---|---|
| 1.0000 | .7500 | -.0001 | -.0000 | -.017 | -.0030 | -.0014 | -.0150 | .0112 | -.0013 | .0003 | .0019 | .0006 | -.0031 | -.0010 |
| 1.2000 | .9000 | -.0002 | -.0000 | -.021 | -.0050 | -.0025 | -.0197 | .0146 | -.0029 | .0009 | .0023 | .0008 | -.0048 | -.0017 |
| 1.4000 | 1.0500 | -.0003 | -.0001 | -.024 | -.0070 | -.0038 | -.0234 | .0173 | -.0051 | .0018 | .0024 | .0008 | -.0064 | -.0024 |
| 1.6000 | 1.2000 | -.0004 | -.0002 | -.026 | -.0087 | -.0050 | -.0262 | .0191 | -.0075 | .0029 | .0022 | .0007 | -.0079 | -.0032 |
| 1.8000 | 1.3500 | -.0006 | -.0002 | -.027 | -.0100 | -.0060 | -.0261 | .0202 | -.0100 | .0041 | .0018 | .0005 | -.0091 | -.0038 |
| 2.0000 | 1.5000 | -.0007 | -.0003 | -.028 | -.0110 | -.0068 | -.0294 | .0208 | -.0124 | .0054 | .0012 | .0003 | -.0102 | -.0044 |
| 3.0000 | 2.2500 | -.0013 | -.0006 | -.029 | -.0130 | -.0086 | -.0306 | .0199 | -.0214 | .0107 | -.0015 | -.0010 | -.0130 | -.0065 |
| 4.0000 | 3.0000 | -.0016 | -.0009 | -.029 | -.0133 | -.0089 | -.0296 | .0180 | -.0256 | .0136 | -.0031 | -.0018 | -.0138 | -.0073 |
| 5.0000 | 3.7500 | -.0017 | -.0009 | -.028 | -.0133 | -.0090 | -.0288 | .0168 | -.0273 | .0149 | -.0038 | -.0022 | -.0140 | -.0077 |

TABLE B.72. NUMERICAL VALUES FOR FACTORS IN EQUS. (3.51) TO (3.56).

SPR......= .500
SPL......= .281
PUL......= 0.000
PDR......= 1.000
POIS.R...= .300
K........= .750

| RR / KL | KWR / KWL | KMC | KQXR / KQXL | KQYRN / KQYLN | KQYRF / KQYLF | KMXR / KMXL | KMYR / KMYL |
|---|---|---|---|---|---|---|---|
| 1.0000 | -.0001 |       | -.0025 | -.0123 | -.0011 | .0021 | -.0031 |
| .7500  | -.0000 | -.017 | -.0011 | -.0092 | .0003  | .0007 | -.0010 |
| 1.2000 | -.0001 |       | -.0041 | -.0162 | -.0024 | .0027 | -.0047 |
| .9000  | -.0000 | -.020 | -.0021 | -.0121 | .0008  | .0009 | -.0017 |
| 1.4000 | -.0002 |       | -.0058 | -.0195 | -.0042 | .0029 | -.0063 |
| 1.0500 | -.0001 | -.023 | -.0031 | -.0144 | .0015  | .0010 | -.0024 |
| 1.6000 | -.0004 |       | -.0073 | -.0220 | -.0063 | .0028 | -.0076 |
| 1.2000 | -.0001 | -.024 | -.0042 | -.0160 | .0024  | .0010 | -.0031 |
| 1.8000 | -.0005 |       | -.0085 | -.0238 | -.0084 | .0025 | -.0087 |
| 1.3500 | -.0002 | -.026 | -.0050 | -.0171 | .0035  | .0009 | -.0037 |
| 2.0000 | -.0006 |       | -.0094 | -.0250 | -.0106 | .0020 | -.0096 |
| 1.5000 | -.0003 | -.026 | -.0058 | -.0177 | .0046  | .0007 | -.0042 |
| 3.0000 | -.0011 |       | -.0113 | -.0268 | -.0186 | .0007 | -.0119 |
| 2.2500 | -.0005 | -.027 | -.0075 | -.0174 | .0093  | -.0005 | -.0059 |
| 4.0000 | -.0014 |       | -.0115 | -.0262 | -.0225 | -.0024 | -.0125 |
| 3.0000 | -.0007 | -.026 | -.0078 | -.0160 | .0119  | -.0014 | -.0066 |
| 5.0000 | -.0015 |       | -.0116 | -.0257 | -.0242 | -.0032 | -.0126 |
| 3.7500 | -.0003 | -.025 | -.0078 | -.0150 | .0132  | -.0018 | -.0069 |

TABLE B.73. NUMERICAL VALUES FOR FACTORS IN EQUS. (3.51) TO (3.56).

```
SPR.......= 0.000
SPL.......= 0.000
POL.......= 1.000
PDR.......= 0.000
POIS.R...= .300
K........= .750
```

| KR RL | KWK KWL | KMC | KQXR KQXL | KQYRN KQYLN | KQYRF KQYLF | KMXR KMXL | KMYR KMYL |
|---|---|---|---|---|---|---|---|
| 1.0000 | -.000H | | -.0348 | .1571 | -.0150 | -.0068 | -.0055 |
| .7500 | -.0002 | -.053 | -.0158 | .1174 | -.0039 | -.0038 | -.0011 |
| 1.2000 | -.0015 | | -.0472 | .1666 | -.0272 | -.0123 | -.0108 |
| .9000 | -.0005 | -.068 | -.0237 | .1236 | .0085 | .0059 | -.0029 |
| 1.4000 | -.0022 | | -.0569 | .1696 | -.0405 | -.0151 | -.0170 |
| 1.0500 | -.0006 | -.080 | -.0306 | .1247 | .0143 | -.0078 | -.0053 |
| 1.6000 | -.0030 | | -.0638 | .1684 | -.0537 | -.0172 | -.0233 |
| 1.2000 | -.0011 | -.091 | -.0361 | .1222 | .0207 | .0093 | -.0082 |
| 1.8000 | -.0037 | | -.0685 | .1651 | -.0658 | -.0187 | -.0294 |
| 1.3500 | -.0015 | -.099 | -.0402 | .1178 | .0271 | -.0105 | -.0113 |
| 2.0000 | -.0344 | | -.0716 | .1607 | -.0765 | -.0196 | -.0349 |
| 1.5000 | -.0016 | -.106 | -.0432 | .1126 | .0333 | -.0114 | -.0142 |
| 3.0000 | -.0066 | | -.0764 | .1405 | -.1097 | -.0206 | -.0528 |
| 2.2500 | -.0032 | -.122 | -.0467 | .0894 | .0553 | .0125 | -.0257 |
| 4.0000 | -.0375 | | -.0768 | .1307 | -.1217 | -.0199 | -.0599 |
| 3.0000 | -.0039 | -.126 | -.0493 | .0776 | .0653 | -.0120 | -.0315 |
| 5.0000 | -.0078 | | -.0768 | .1272 | -.1255 | -.0194 | -.0624 |
| 3.7500 | -.0043 | -.127 | -.0444 | .0729 | .0693 | -.0114 | -.0341 |

TABLE B.74. NUMERICAL VALUES FOR FACTORS IN EQUS. (3.51) TO (3.56).

```
SPR.......=  .100
SPL.......=  .056
POL.......= 1.000
POR.......= 0.000
PUIS.R...=  .300
K........=  .750
```

| RR / RL | KWR / KWL | KMC | KQXR / KQXL | KQYRN / KQYLN | KQYRF / KQYLF | KMXR / KMXL | KMYR / KMYL |
|---|---|---|---|---|---|---|---|
| 1.0000 | -.0003 |       | -.0115 | -.0570 | -.0050 | -.0004 | -.0043 |
| .7500  | -.0001 | -.030 | -.0053 |  .0426 |  .0013 | -.0004 | -.0012 |
| 1.2000 | -.0005 |       | -.0193 | -.0756 | -.0112 | -.0016 | -.0080 |
| .9000  | -.0002 | -.042 | -.0097 |  .0562 |  .0035 | -.0011 | -.0025 |
| 1.4000 | -.0011 |       | -.0269 | -.0904 | -.0195 | -.0032 | -.0123 |
| 1.0500 | -.0004 | -.053 | -.0146 |  .0666 |  .0069 | -.0021 | -.0043 |
| 1.6000 | -.0016 |       | -.0334 | -.1011 | -.0249 | -.0049 | -.0170 |
| 1.2000 | -.0006 | -.063 | -.0191 |  .0736 |  .0111 | -.0031 | -.0064 |
| 1.8000 | -.0022 |       | -.0385 | -.1083 | -.0385 | -.0066 | -.0216 |
| 1.3500 | -.0009 | -.071 | -.0230 |  .0778 |  .0158 | -.0041 | -.0086 |
| 2.0000 | -.0026 |       | -.0423 | -.1128 | -.0476 | -.0082 | -.0260 |
| 1.5000 | -.0011 | -.078 | -.0260 |  .0797 |  .0206 | -.0051 | -.0108 |
| 3.0000 | -.0049 |       | -.0492 | -.1152 | -.0806 | -.0132 | -.0412 |
| 2.2500 | -.0024 | -.097 | -.0326 |  .0749 |  .0402 | -.0080 | -.0201 |
| 4.0000 | -.0056 |       | -.0499 | -.1097 | -.0955 | -.0150 | -.0480 |
| 3.0000 | -.0631 | -.102 | -.0336 |  .0666 |  .0507 | -.0089 | -.0251 |
| 5.0000 | -.0063 |       | -.0500 | -.1060 | -.1011 | -.0155 | -.0505 |
| 3.7500 | -.0034 | -.103 | -.0337 |  .0617 |  .0554 | -.0090 | -.0275 |

TABLE B.75. NUMERICAL VALUES FOR FACTORS IN EQUS. (3.51) TO (3.56).

SPR...... = .200
SPL...... = .113
PGL...... = 1.000
POR...... = 0.000
PUIS.P... = .300
K........ = .750

| RR / RL | KWR / KWL | KMC | KQXR / KQXL | KQYRN / KQYLN | KQYRF / KQYLF | KMXR / KMXL | KMYR / KMYL |
|---|---|---|---|---|---|---|---|
| 1.0000 | -.0002 | | -.0069 | -.0343 | -.0030 | .0013 | -.0041 |
| .7500 | -.0000 | -.026 | -.0032 | -.0256 | .0008 | .0003 | -.0013 |
| 1.2000 | -.0004 | | -.0122 | -.0478 | -.0071 | .0012 | -.0072 |
| .9000 | -.0001 | -.035 | -.0061 | -.0356 | .0022 | .0001 | -.0024 |
| 1.4000 | -.0007 | | -.0177 | -.0599 | -.0129 | -.0006 | -.0109 |
| 1.0500 | -.0002 | -.044 | -.0096 | -.0441 | .0045 | -.0002 | -.0039 |
| 1.6000 | -.0011 | | -.0228 | -.0697 | -.0198 | -.0003 | -.0147 |
| 1.2000 | -.0004 | -.052 | -.0131 | -.0508 | .0076 | -.0008 | -.0057 |
| 1.8000 | -.0015 | | -.0271 | -.0773 | -.0272 | -.0015 | -.0185 |
| 1.3500 | -.0006 | -.059 | -.0162 | -.0555 | .0112 | -.0015 | -.0075 |
| 2.0000 | -.0020 | | -.0305 | -.0828 | -.0345 | -.0027 | -.0220 |
| 1.5000 | -.0008 | -.065 | -.0188 | -.0586 | .0149 | -.0022 | -.0094 |
| 3.0000 | -.0036 | | -.0377 | -.0928 | -.0632 | -.0082 | -.0345 |
| 2.2500 | -.0018 | -.081 | -.0251 | -.0605 | .0315 | -.0051 | -.0169 |
| 4.0000 | -.0048 | | -.0387 | -.0918 | -.0778 | -.0112 | -.0402 |
| 3.0000 | -.0025 | -.086 | -.0263 | -.0561 | .0412 | -.0066 | -.0211 |
| 5.0000 | -.0052 | | -.0388 | -.0897 | -.0840 | -.0125 | -.0425 |
| 3.7500 | -.0026 | -.087 | -.0264 | -.0526 | .0459 | -.0072 | -.0231 |

TABLE B.76. NUMERICAL VALUES FOR FACTORS IN EQUS. (3.51) TO (3.56).

```
SPR...... =  .300
SPL...... =  .169
PÜI...... = 1.000
PÜR...... = 0.000
POIS.R... =  .300
K........ =  .750
```

| KR / RL | KRR / KRL | KMC | KQXR / KQXL | KQYRN / KQYLN | KQYRF / KQYLF | KMXR / KMXL | KMYR / KMYL |
|---|---|---|---|---|---|---|---|
| 1.0000 | -.0001 |  | -.0049 | -.0245 | -.0021 | .0020 | -.0040 |
| .7500 | -.0000 | -.024 | -.0023 | .0183 | .0006 | .0006 | -.0013 |
| 1.2000 | -.0003 |  | -.0069 | -.0350 | -.0052 | .0025 | -.0069 |
| .9000 | -.0001 | -.032 | -.0045 | .0260 | .0016 | .0007 | -.0024 |
| 1.4000 | -.0005 |  | -.0132 | -.0447 | -.0096 | .0025 | -.0102 |
| 1.0500 | -.0002 | -.040 | -.0072 | .0330 | .0034 | .0007 | -.0038 |
| 1.6000 | -.0006 |  | -.0173 | -.0531 | -.0150 | .0021 | -.0135 |
| 1.2000 | -.0003 | -.046 | -.0099 | .0387 | .0058 | .0004 | -.0053 |
| 1.8000 | -.0012 |  | -.0209 | -.0599 | -.0210 | .0013 | -.0168 |
| 1.3500 | -.0005 | -.052 | -.0125 | .0430 | .0086 | .0000 | -.0069 |
| 2.0000 | -.0016 |  | -.0238 | -.0653 | -.0270 | .0004 | -.0198 |
| 1.5000 | -.0006 | -.057 | -.0147 | .0462 | .0117 | -.0005 | -.0085 |
| 3.0000 | -.0031 |  | -.0306 | -.0771 | -.0520 | -.0049 | -.0302 |
| 2.2500 | -.0015 | -.071 | -.0205 | .0504 | .0256 | -.0032 | -.0149 |
| 4.0000 | -.0040 |  | -.0318 | -.0783 | -.0655 | -.0084 | -.0349 |
| 3.0000 | -.0021 | -.075 | -.0217 | .0480 | .0346 | -.0050 | -.0183 |
| 5.0000 | -.0044 |  | -.0319 | -.0774 | -.0717 | -.0102 | -.0368 |
| 3.7500 | -.0024 | -.076 | -.0219 | .0455 | .0391 | -.0058 | -.0200 |

TABLE B.77. NUMERICAL VALUES FOR FACTORS IN EQUS. (3.51) TO (3.56).

SFR......... = .400
SPL......... = .225
POL......... = 1.000
POR......... = 0.000
PUIS.R..... = .300
K.......... = .750

| RK KL | KWR KWL | KMC | KQXR KQXL | KQYRN KQYLN | KQYKF KQYLF | KMXR KMXL | KMYR KMYL |
|---|---|---|---|---|---|---|---|
| 1.0000 | -.0001 | | -.0038 | -.0191 | -.0017 | .0024 | -.0040 |
| .7500 | -.0000 | -.022 | -.0018 | .0143 | -.0004 | .0007 | -.0013 |
| 1.2000 | -.0002 | | -.0070 | -.0276 | -.0041 | .0032 | -.0067 |
| .9000 | -.0001 | -.030 | -.0035 | -.0205 | -.0013 | .0011 | -.0024 |
| 1.4000 | -.0004 | | -.0105 | -.0357 | -.0076 | .0037 | -.0097 |
| 1.0500 | -.0001 | -.037 | -.0657 | .0263 | .0027 | .0012 | -.0037 |
| 1.6000 | -.0007 | | -.0140 | -.0429 | -.0121 | .0036 | -.0126 |
| 1.2000 | -.0003 | -.043 | -.0080 | -.0312 | -.0047 | .0012 | -.0051 |
| 1.8000 | -.0010 | | -.0170 | -.0489 | -.0171 | .0031 | -.0157 |
| 1.3500 | -.0004 | -.048 | -.0102 | -.0351 | -.0070 | .0009 | -.0066 |
| 2.0000 | -.0013 | | -.0196 | -.0538 | -.0222 | .0024 | -.0183 |
| 1.5000 | -.0005 | -.052 | -.0121 | -.0381 | -.0096 | .0005 | -.0080 |
| 3.0000 | -.0026 | | -.0258 | -.0658 | -.0441 | -.0026 | -.0272 |
| 2.2500 | -.0013 | -.063 | -.0173 | -.0431 | .0219 | -.0019 | -.0135 |
| 4.0000 | -.0035 | | -.0271 | -.0681 | -.0565 | -.0063 | -.0311 |
| 3.0000 | -.0018 | -.066 | -.0185 | .0419 | .0298 | -.0038 | -.0163 |
| 5.0000 | -.0039 | | -.0272 | -.0679 | -.0625 | -.0084 | -.0326 |
| 3.7500 | -.0021 | -.067 | -.0187 | .0401 | .0340 | -.0048 | -.0177 |

TABLE B.78. NUMERICAL VALUES FOR FACTORS IN EQUS. (3.51) TO (3.56).

```
SPR.......=   .500
SPL.......=   .281
POL.......= 1.000
POR.......= 0.000
POIS.R...=   .300
K.........=   .750
```

| KR / RL | KWR / KWL | KMC | KQXK / KQXL | KQYRN / KQYLN | KQYRF / KQYLF | KMXR / KMXL | KMYR / KMYL |
|---|---|---|---|---|---|---|---|
| 1.0000 | -.0001 |        | -.0031 | -.0156 | -.0014 | .0026 | -.0039 |
|  .7500 | -.0000 | -.022  | -.0014 |  .0117 |  .0004 | .0008 | -.0013 |
| 1.2000 | -.0002 |        | -.0058 | -.0227 | -.0034 | .0037 | -.0066 |
|  .9000 | -.0001 | -.029  | -.0029 |  .0169 |  .0010 | .0013 | -.0023 |
| 1.4000 | -.0003 |        | -.0087 | -.0297 | -.0064 | .0044 | -.0095 |
| 1.0500 | -.0001 | -.035  | -.0047 |  .0219 |  .0022 | .0016 | -.0036 |
| 1.6000 | -.0006 |        | -.0117 | -.0360 | -.0101 | .0046 | -.0123 |
| 1.2000 | -.0002 | -.041  | -.0067 |  .0262 |  .0039 | .0017 | -.0050 |
| 1.8000 | -.0006 |        | -.0144 | -.0413 | -.0144 | .0043 | -.0149 |
| 1.3500 | -.0003 | -.045  | -.0086 |  .0297 |  .0059 | .0015 | -.0063 |
| 2.0000 | -.0011 |        | -.0166 | -.0458 | -.0189 | .0038 | -.0173 |
| 1.5000 | -.0005 | -.049  | -.0103 |  .0324 |  .0062 | .0013 | -.0076 |
| 3.0000 | -.0623 |        | -.0223 | -.0574 | -.0383 | -.0009 | -.0250 |
| 2.2500 | -.0011 | -.058  | -.0150 |  .0376 |  .0190 | -.0009 | -.0124 |
| 4.0000 | -.0030 |        | -.0236 | -.0602 | -.0497 | -.0048 | -.0282 |
| 3.0000 | -.0016 | -.060  | -.0162 |  .0371 |  .0262 | -.0028 | -.0146 |
| 5.0000 | -.0034 |        | -.0238 | -.0604 | -.0553 | -.0070 | -.0293 |
| 3.7500 | -.0018 | -.060  | -.0164 |  .0357 |  .0301 | -.0040 | -.0159 |

TABLE B.79. NUMERICAL VALUES FOR FACTORS IN EQUS. (3.51) TO (3.56).

```
SPP......= 0.000
SPL......= 0.000
POI......= 0.000
PUR......= 1.000
POIS.R...= .300
K........= 1.000
```

| I KK I | KKK I | KMC I | KQXR I | KQYRN I | KQYRF I | KMXR I | KMYR I |
| I RL I | KWL I |   I | KQXL I | KQYLN I | KQYLF I | KMXL I | KMYL I |
|---|---|---|---|---|---|---|---|
| 1.0000 | -.0006 |  | -.0277 | -.1214 | -.0119 | -.0070 | -.0044 |
| 1.0000 | -.0006 | -.042 | -.0277 | .1214 | .0119 | -.0070 | -.0044 |
| 1.2000 | -.0011 |  | -.0347 | -.1165 | -.0199 | -.0090 | -.0079 |
| 1.2000 | -.0011 | -.049 | -.0347 | .1165 | .0199 | -.0090 | -.0079 |
| 1.4000 | -.0015 |  | -.0392 | -.1087 | -.0277 | -.0103 | -.0116 |
| 1.4000 | -.0015 | -.054 | -.0392 | .1087 | .0277 | -.0103 | -.0116 |
| 1.6000 | -.0019 |  | -.0419 | -.1003 | -.0347 | -.0110 | -.0151 |
| 1.6000 | -.0019 | -.057 | -.0419 | .1003 | .0347 | -.0110 | -.0151 |
| 1.8000 | -.0023 |  | -.0434 | -.0925 | -.0406 | -.0113 | -.0182 |
| 1.8000 | -.0023 | -.059 | -.0434 | .0925 | .0406 | -.0113 | -.0182 |
| 2.0000 | -.0026 |  | -.0443 | -.0857 | -.0455 | -.0114 | -.0208 |
| 2.0000 | -.0026 | -.061 | -.0443 | .0857 | .0455 | -.0114 | -.0208 |
| 3.0000 | -.0035 |  | -.0452 | -.0673 | -.0583 | -.0106 | -.0282 |
| 3.0000 | -.0036 | -.062 | -.0452 | .0673 | .0583 | -.0106 | -.0282 |
| 4.0000 | -.0038 |  | -.0452 | -.0626 | -.0616 | -.0098 | -.0305 |
| 4.0000 | -.0038 | -.062 | -.0452 | .0626 | .0616 | -.0098 | -.0305 |
| 5.0000 | -.0039 |  | -.0452 | -.0614 | -.0623 | -.0095 | -.0311 |
| 5.0000 | -.0039 | -.062 | -.0452 | .0614 | .0623 | -.0095 | -.0311 |

TABLE B.80. NUMERICAL VALUES FOR FACTORS IN EQUS. (3.51) TO (3.56).

```
SPR.......=  .100
SPL.......=  .100
POL.......= 0.000
POR.......= 1.000
POIS.K...=  .300
K.........= 1.000
```

| RR / RL | KWR / KWL | KMC | KQXK / KQXL | KQYRN / KQYLN | KQYRF / KQYLF | KMXR / KMXL | KMYR / KMYL |
|---|---|---|---|---|---|---|---|
| 1.0000 | -.0002 |       | -.0091 | -.0448 | -.0040 | -.0003 | -.0034 |
| 1.0000 | -.0002 | -.024 | -.0091 |  .0448 |  .0040 | -.0003 | -.0034 |
| 1.2000 | -.0004 |       | -.0140 | -.0541 | -.0081 | -.0012 | -.0057 |
| 1.2000 | -.0004 | -.030 | -.0140 |  .0541 |  .0081 | -.0012 | -.0057 |
| 1.4000 | -.0007 |       | -.0181 | -.0597 | -.0131 | -.0021 | -.0062 |
| 1.4000 | -.0007 | -.035 | -.0181 |  .0597 |  .0131 | -.0021 | -.0082 |
| 1.6000 | -.0010 |       | -.0211 | -.0623 | -.0182 | -.0031 | -.0107 |
| 1.6000 | -.0010 | -.039 | -.0211 |  .0623 |  .0182 | -.0031 | -.0107 |
| 1.8000 | -.0013 |       | -.0233 | -.0631 | -.0230 | -.0040 | -.0129 |
| 1.8000 | -.0013 | -.042 | -.0233 |  .0631 |  .0230 | -.0040 | -.0129 |
| 2.0000 | -.0016 |       | -.0246 | -.0626 | -.0274 | -.0047 | -.0149 |
| 2.0000 | -.0016 | -.044 | -.0246 |  .0626 |  .0274 | -.0047 | -.0149 |
| 3.0000 | -.0025 |       | -.0266 | -.0554 | -.0412 | -.0068 | -.0209 |
| 3.0000 | -.0025 | -.048 | -.0266 |  .0554 |  .0412 | -.0068 | -.0209 |
| 4.0000 | -.0028 |       | -.0267 | -.0505 | -.0462 | -.0073 | -.0231 |
| 4.0000 | -.0028 | -.048 | -.0267 |  .0505 |  .0462 | -.0073 | -.0231 |
| 5.0000 | -.0030 |       | -.0267 | -.0487 | -.0476 | -.0073 | -.0238 |
| 5.0000 | -.0030 | -.048 | -.0267 |  .0487 |  .0476 | -.0073 | -.0238 |

TABLE B.81. NUMERICAL VALUES FOR FACTORS IN EQUS. (3.51) TO (3.56).

```
SPR...... =  .200
SPL...... =  .200
POL...... = 0.000
POR...... = 1.000
POIS.R... =  .300
K........ = 1.000
```

| RR RL | KWR KWL | KMC | KQXR KQXL | KQYRN KQYLN | KQYRF KQYLF | KMXR KMXL | KMYR KMYL |
|---|---|---|---|---|---|---|---|
| 1.0000 | −.0001 |        | −.0055 | −.0269 | −.0024 |  .0010 | −.0033 |
| 1.0000 | −.0001 | −.020 | −.0055 |  .0269 |  .0024 |  .0010 | −.0033 |
| 1.2000 | −.0003 |        | −.0088 | −.0342 | −.0051 |  .0009 | −.0052 |
| 1.2000 | −.0003 | −.025 | −.0088 |  .0342 |  .0051 |  .0009 | −.0052 |
| 1.4000 | −.0005 |        | −.0118 | −.0394 | −.0086 |  .0004 | −.0072 |
| 1.4000 | −.0005 | −.029 | −.0118 |  .0394 |  .0086 |  .0004 | −.0072 |
| 1.6000 | −.0007 |        | −.0143 | −.0427 | −.0123 | −.0002 | −.0091 |
| 1.6000 | −.0007 | −.032 | −.0143 |  .0427 |  .0123 | −.0002 | −.0091 |
| 1.8000 | −.0009 |        | −.0161 | −.0447 | −.0161 | −.0009 | −.0109 |
| 1.8000 | −.0009 | −.034 | −.0161 |  .0447 |  .0161 | −.0009 | −.0109 |
| 2.0000 | −.0011 |        | −.0174 | −.0456 | −.0195 | −.0016 | −.0124 |
| 2.0000 | −.0011 | −.036 | −.0174 |  .0456 |  .0195 | −.0016 | −.0124 |
| 3.0000 | −.0019 |        | −.0197 | −.0440 | −.0315 | −.0042 | −.0170 |
| 3.0000 | −.0019 | −.039 | −.0197 |  .0440 |  .0315 | −.0042 | −.0170 |
| 4.0000 | −.0022 |        | −.0199 | −.0412 | −.0365 | −.0054 | −.0187 |
| 4.0000 | −.0022 | −.039 | −.0199 |  .0412 |  .0365 | −.0054 | −.0187 |
| 5.0000 | −.0024 |        | −.0199 | −.0398 | −.0383 | −.0057 | −.0193 |
| 5.0000 | −.0024 | −.039 | −.0199 |  .0398 |  .0383 | −.0057 | −.0193 |

TABLE B.82. NUMERICAL VALUES FOR FACTORS IN EQUS. (3.51) TO (3.56).

```
SPR.......=   .300
SPL.......=   .300
POL.......= 0.000
PUR.......= 1.000
POIS.R...=   .300
K.........= 1.000
```

| RP / KL | KWR / KWL | KMC | KQXR / KQXL | KQYRN / KQYLN | KQYRF / KQYLF | KMXR / KMXL | KMYR / KMYL |
|---|---|---|---|---|---|---|---|
| 1.0000 | -.0001 |       | -.0039 | -.0192 | -.0017 |  .0016 | -.0032 |
| 1.0000 | -.0001 | -.018 | -.0039 |  .0192 |  .0017 |  .0016 | -.0032 |
| 1.2000 | -.0002 |       | -.0064 | -.0250 | -.0037 |  .0018 | -.0050 |
| 1.2000 | -.0002 | -.023 | -.0064 |  .0250 |  .0037 |  .0018 | -.0050 |
| 1.4000 | -.0004 |       | -.0088 | -.0294 | -.0064 |  .0017 | -.0067 |
| 1.4000 | -.0004 | -.026 | -.0088 |  .0294 |  .0064 |  .0017 | -.0067 |
| 1.6000 | -.0005 |       | -.0108 | -.0325 | -.0093 |  .0013 | -.0083 |
| 1.6000 | -.0005 | -.028 | -.0108 |  .0325 |  .0093 |  .0013 | -.0083 |
| 1.8000 | -.0007 |       | -.0123 | -.0345 | -.0123 |  .0007 | -.0096 |
| 1.8000 | -.0007 | -.030 | -.0123 |  .0345 |  .0123 |  .0007 | -.0098 |
| 2.0000 | -.0009 |       | -.0135 | -.0357 | -.0152 |  .0001 | -.0110 |
| 2.0000 | -.0009 | -.031 | -.0135 |  .0357 |  .0152 |  .0001 | -.011C |
| 3.0000 | -.0015 |       | -.0157 | -.0362 | -.0255 | -.0026 | -.0146 |
| 3.0000 | -.0015 | -.033 | -.0157 |  .0362 |  .0255 | -.0026 | -.0146 |
| 4.0000 | -.0019 |       | -.0159 | -.0346 | -.0302 | -.0041 | -.0158 |
| 4.0000 | -.0019 | -.033 | -.0159 |  .0346 |  .0302 | -.0041 | -.0158 |
| 5.0000 | -.0020 |       | -.0159 | -.0336 | -.0320 | -.0046 | -.0163 |
| 5.0000 | -.0020 | -.033 | -.0159 |  .0336 |  .0320 | -.0046 | -.0163 |

TABLE B.83. NUMERICAL VALUES FOR FACTORS IN EQUS. (3.51) TO (3.56).

```
SPR......=    .400
SPL......=    .400
POL......=   0.000
PUR......=   1.000
POIS.R...=    .300
K........=   1.000
```

| KR / RL | KWR / KWL | KMC | KQXK / KQXL | KQYRN / KQYLN | KQYRF / KQYLF | KMXR / KMXL | KMYR / KMYL |
|---|---|---|---|---|---|---|---|
| 1.0000 | -.0001 |       | -.0030 | -.0150 | -.0013 | .0019 | -.0031 |
| 1.0000 | -.0001 | -.017 | -.0030 | -.0150 | -.0013 | .0019 | -.0031 |
| 1.2000 | -.0002 |       | -.0050 | -.0197 | -.0029 | .0023 | -.0048 |
| 1.2000 | -.0002 | -.021 | -.0050 | -.0197 | -.0029 | .0023 | -.0048 |
| 1.4000 | -.0003 |       | -.0070 | -.0234 | -.0051 | .0024 | -.0064 |
| 1.4000 | -.0003 | -.024 | -.0070 | -.0234 | -.0051 | .0024 | -.0064 |
| 1.6000 | -.0004 |       | -.0087 | -.0262 | -.0075 | .0022 | -.0079 |
| 1.6000 | -.0004 | -.026 | -.0067 | -.0262 | -.0075 | .0022 | -.0079 |
| 1.8000 | -.0006 |       | -.0100 | -.0281 | -.0100 | .0018 | -.0091 |
| 1.8000 | -.0006 | -.027 | -.0100 | -.0281 | -.0100 | .0018 | -.0091 |
| 2.0000 | -.0007 |       | -.0110 | -.0293 | -.0124 | .0012 | -.0101 |
| 2.0000 | -.0007 | -.028 | -.0110 | -.0293 | -.0124 | .0012 | -.0101 |
| 3.0000 | -.0013 |       | -.0130 | -.0306 | -.0214 | -.0015 | -.0130 |
| 3.0000 | -.0013 | -.029 | -.0130 | -.0306 | -.0214 | -.0015 | -.0130 |
| 4.0000 | -.0016 |       | -.0133 | -.0297 | -.0257 | -.0031 | -.0138 |
| 4.0000 | -.0016 | -.029 | -.0133 | -.0297 | -.0257 | -.0031 | -.0138 |
| 5.0000 | -.0017 |       | -.0133 | -.0290 | -.0274 | -.0038 | -.0141 |
| 5.0000 | -.0017 | -.029 | -.0133 | -.0290 | -.0274 | -.0038 | -.0141 |

TABLE B.84. NUMERICAL VALUES FOR FACTORS IN EQUS. (3.51) TO (3.56)

345

SPR........ = .500
SPL........ = .500
POL........ = 0.000
PUR........ = 1.000
POIS.R... = .300
K......... = 1.000

| KR / FL | KWR / KWL | KMC | KQXR / KQXL | KQYRN / KQYLN | KQYRF / KQYLF | KMXR / KMXL | KMYR / KMYL |
|---|---|---|---|---|---|---|---|
| 1.0000 | -.0001 |       | -.0025 | -.0122 | -.0011 | .0021 | -.0031 |
| 1.0000 | -.0001 | -.017 | -.0025 | .0122 | .0011 | .0021 | -.0031 |
| 1.2000 | -.0001 |       | -.0041 | -.0162 | -.0024 | .0027 | -.0047 |
| 1.2000 | -.0001 | -.020 | -.0041 | .0162 | .0024 | .0027 | -.0047 |
| 1.4000 | -.0002 |       | -.0058 | -.0194 | -.0042 | .0029 | -.0062 |
| 1.4000 | -.0002 | -.023 | -.0058 | .0194 | .0042 | .0029 | -.0062 |
| 1.6000 | -.0004 |       | -.0072 | -.0219 | -.0063 | .0028 | -.0076 |
| 1.6000 | -.0004 | -.024 | -.0072 | .0219 | .0063 | .0028 | -.0076 |
| 1.8000 | -.0005 |       | -.0084 | -.0237 | -.0084 | .0025 | -.0086 |
| 1.8000 | -.0005 | -.025 | -.0084 | .0237 | .0084 | .0025 | -.0086 |
| 2.0000 | -.0006 |       | -.0093 | -.0249 | -.0105 | .0020 | -.0095 |
| 2.0000 | -.0006 | -.026 | -.0093 | .0249 | .0105 | .0020 | -.0095 |
| 3.0000 | -.0011 |       | -.0112 | -.0265 | -.0184 | -.0007 | -.0118 |
| 3.0000 | -.0011 | -.026 | -.0112 | .0265 | .0184 | -.0007 | -.0118 |
| 4.0000 | -.0014 |       | -.0115 | -.0260 | -.0223 | -.0024 | -.0124 |
| 4.0000 | -.0014 | -.026 | -.0115 | .0260 | .0223 | -.0024 | -.0124 |
| 5.0000 | -.0015 |       | -.0115 | -.0255 | -.0240 | -.0032 | -.0125 |
| 5.0000 | -.0015 | -.025 | -.0115 | .0255 | .0240 | -.0032 | -.0125 |

TABLE B.85. NUMERICAL VALUES FOR FACTORS IN EQUS. (3.51) TO (3.56).

# APPENDIX C
## Explicit Forms of the Matrices Used in the Finite Analysis

The strain-displacement matrix is defined by Equ. (6.19) as

$$[B] \quad = \quad [B_1 \quad B_2 \quad B_3 \quad B_4 \quad B_5 \quad B_6 \quad B_7 \quad B_8] \tag{6.19}$$

in which

$$[B_1] = \begin{bmatrix} \frac{1}{8a}(1+\eta)(1+\zeta) & 0 & 0 \\ 0 & \frac{1}{8b}(1+\xi)(1+\zeta) & 0 \\ 0 & 0 & \frac{1}{8c}(1+\xi)(1+\eta) \\ \frac{1}{8b}(1+\xi)(1+\zeta) & \frac{1}{8a}(1+\eta)(1+\zeta) & 0 \\ 0 & \frac{1}{8c}(1+\xi)(1+\eta) & \frac{1}{8b}(1+\xi)(1+\zeta) \\ \frac{1}{8c}(1+\xi)(1+\eta) & 0 & \frac{1}{8a}(1+\eta)(1+\zeta) \end{bmatrix}$$

$$[B_2] = \begin{bmatrix} \frac{1}{8a}(1-n)(1+\varsigma) & 0 & 0 \\[6pt] 0 & -\frac{1}{8b}(1+\xi)(1+\varsigma) & 0 \\[6pt] 0 & 0 & \frac{1}{8c}(1+\xi)(1-n) \\[6pt] -\frac{1}{8b}(1+\xi)(1+\varsigma) & \frac{1}{8a}(1-n)(1+\varsigma) & 0 \\[6pt] 0 & \frac{1}{8c}(1+\xi)(1-n) & -\frac{1}{8b}(1+\xi)(1+\varsigma) \\[6pt] \frac{1}{8c}(1+\xi)(1-n) & 0 & \frac{1}{8a}(1-n)(1+\varsigma) \end{bmatrix}$$

$$[B_3] = \begin{bmatrix} -\frac{1}{8a}(1-n)(1+\varsigma) & 0 & 0 \\[6pt] 0 & -\frac{1}{8b}(1-\xi)(1+\varsigma) & 0 \\[6pt] 0 & 0 & \frac{1}{8c}(1-\xi)(1-n) \\[6pt] -\frac{1}{8b}(1-\xi)(1+\varsigma) & -\frac{1}{8a}(1-n)(1+\varsigma) & 0 \\[6pt] 0 & \frac{1}{8c}(1-\xi)(1+n) & -\frac{1}{8b}(1-\xi)(1+\varsigma) \\[6pt] \frac{1}{8c}(1-\xi)(1+n) & 0 & -\frac{1}{8a}(1-n)(1+\varsigma) \end{bmatrix}$$

$$[B_4] = \begin{bmatrix} -\dfrac{1}{8a}(1+n)(1+\zeta) & 0 & 0 \\[2em] 0 & \dfrac{1}{8b}(1-\xi)(1+\zeta) & 0 \\[2em] 0 & 0 & \dfrac{1}{8c}(1-\xi)(1+n) \\[2em] \dfrac{1}{8b}(1-\xi)(1+\zeta) & -\dfrac{1}{8a}(1+n)(1+\zeta) & 0 \\[2em] 0 & \dfrac{1}{8c}(1-\xi)(1+n) & \dfrac{1}{8b}(1-\xi)(1+\zeta) \\[2em] \dfrac{1}{8c}(1-\xi)(1+n) & 0 & -\dfrac{1}{8a}(1+n)(1+\zeta) \end{bmatrix}$$

$$[B_5] = \begin{bmatrix} \dfrac{1}{8a}(1+n)(1-\zeta) & 0 & 0 \\[2em] 0 & \dfrac{1}{8b}(1+\xi)(1-\zeta) & 0 \\[2em] 0 & 0 & -\dfrac{1}{8c}(1+\xi)(1+n) \\[2em] \dfrac{1}{8b}(1+\xi)(1-\zeta) & \dfrac{1}{8a}(1+n)(1-\zeta) & 0 \\[2em] 0 & -\dfrac{1}{8c}(1+\xi)(1+n) & \dfrac{1}{8b}(1+\xi)(1-\zeta) \\[2em] -\dfrac{1}{8c}(1+\xi)(1+n) & 0 & \dfrac{1}{8a}(1+n)(1-\zeta) \end{bmatrix}$$

$$[B_6] = \begin{bmatrix}
\frac{1}{8a}(1-n)(1-\zeta) & 0 & 0 \\
0 & -\frac{1}{8b}(1+\xi)(1-\zeta) & 0 \\
0 & 0 & -\frac{1}{8c}(1+\xi)(1-n) \\
-\frac{1}{8b}(1+\xi)(1-\zeta) & \frac{1}{8a}(1-n)(1-\zeta) & 0 \\
0 & -\frac{1}{8c}(1+\xi)(1-n) & -\frac{1}{8b}(1+\xi)(1-\zeta) \\
-\frac{1}{8c}(1+\xi)(1-n) & 0 & \frac{1}{8a}(1-n)(1-\zeta)
\end{bmatrix}$$

$$[B_7] = \begin{bmatrix}
-\frac{1}{8a}(1-n)(1-\zeta) & 0 & 0 \\
0 & -\frac{1}{8b}(1-\xi)(1-\zeta) & 0 \\
0 & 0 & -\frac{1}{8c}(1-\xi)(1-n) \\
-\frac{1}{8b}(1-\xi)(1-\zeta) & -\frac{1}{8a}(1-n)(1-\zeta) & 0 \\
0 & -\frac{1}{8c}(1-\xi)(1-n) & -\frac{1}{8b}(1-\xi)(1-\zeta) \\
-\frac{1}{8c}(1-\xi)(1-n) & 0 & -\frac{1}{8a}(1-n)(1-\zeta)
\end{bmatrix}$$

$$[B_8] = \begin{bmatrix} -\dfrac{1}{8a}(1+\eta)(1-\zeta) & 0 & 0 \\[2em] 0 & \dfrac{1}{8b}(1-\xi)(1-\zeta) & 0 \\[2em] 0 & 0 & -\dfrac{1}{8c}(1-\xi)(1+\eta) \\[2em] \dfrac{1}{8b}(1-\xi)(1-\zeta) & -\dfrac{1}{8a}(1+\eta)(1-\zeta) & 0 \\[2em] 0 & -\dfrac{1}{8c}(1-\xi)(1+\eta) & \dfrac{1}{8b}(1-\xi)(1-\zeta) \\[2em] -\dfrac{1}{8c}(1-\xi)(1+\eta) & 0 & -\dfrac{1}{8a}(1+\eta)(1-\zeta) \end{bmatrix}$$

# INDEX